PERFORMANCE ASSESSMENT
AND ENRICHMENT OF
ANAEROBIC METHANE OXIDISING
MICROBIAL COMMUNITIES
FROM MARINE SEDIMENTS
IN BIOREACTORS

Susma Bhattarai Gautam

Joint PhD degree in Environmental Technology

UNIVERSITÉ ━━━━
━ PARIS-EST

Docteur de l'Université Paris-Est
Spécialité : Science et Technique de l'Environnement

Dottore di Ricerca in Tecnologie Ambientali

UNESCO-IHE
Institute for Water Education

Degree of Doctor in Environmental Technology

Thèse - Tesi di Dottorato - PhD thesis

Susma Bhattarai Gautam

Performance Assessment and Enrichment of Anaerobic Methane
Oxidizing Microbial Communities from Marine Sediments in Bioreactors

Defended on December 16th, 2016

In front of the PhD committee

Dr. Cartsen Vogt	Reviewer
Dr. Artin Hatzikioseyian	Reviewer
Prof. dr. ir. Piet. N. L. Lens	Promotor
Prof. Michel Madon	Co-Promotor
Prof. Giovanni Esposito	Co-Promotor
Dr. Eldon R. Rene	Co-Promotor
Hab. Dr. Eric van Hullebusch	Examiner

Erasmus Joint doctorate programme in Environmental Technology for Contaminated Solids, Soils and
Sediments (ETeCoS³)

Thesis Committee

Thesis Promoter

Prof. dr. ir. Piet N. L. Lens
Professor of Environmental Biotechnology
UNESCO-IHE, Institute for Water Education
Delft, the Netherlands

Thesis Co-Promoters

Prof. Michel Madon
University of Paris-Est, Institut Francilien des Sciences Appliqueées
Champs sur Marne, France

Dr. Giovanni Esposito
Associate Professor of Sanitary and Environmental Engineering
University of Cassino and Southern Lazio, Cassino, Italy

Dr. Eldon R. Rene
UNESCO-IHE, Institute for Water Education
Delft, the Netherlands

Examiner

Hab. Dr. Eric van Hullebusch
Associate Professor of Biogeochemistry
University of Paris-Est, Institut Francilien des Sciences Appliqueées
Champs sur Marne, France

Reviewers

Dr. Cartsen Vogt
Helmholtz Centre for Environmental Research - UFZ
Department of Isotope Biogeochemistry
Leipzig, Germany

Dr. Artin Hatzikioseyian
National Technical University of Athens
Athens, Greece

This research was conducted under the auspices of the Erasmus Mundus Joint Doctorate Environmental Technologies for Contaminated Solids, Soils, and Sediments (ETeCoS[3]) and The Netherlands Research School for the Socio-Economic and Natural Sciences of the Environment (SENSE).

CRC Press/Balkema is an imprint of the Taylor & Francis Group, an informa business

© 2018, Susma Bhattarai Gautam

Published by: CRC Press/Balkema
PO Box 11320, 2301 EH Leiden, The Netherlands
e-mail: Pub.NL@taylorandfrancis.com
www.crcpress.com – www.taylorandfrancis.com

ISBN: 978-1-138-33021-4 (Taylor & Francis Group)

Table of contents

Acknowledgements

I would like to express my sincere gratitude to my supervisor and PhD promotor, Prof. Piet N. L. Lens, for his priceless suggestions, inspirations and valuable supports throughout the research and this thesis.

I would like to acknowledge my PhD co-promoter, Dr. Giovanni Esposito for his supports and inspirations during this research. Futher extended acknowledgement to my mentor and my thesis co-promoter. Dr. Eldon R. Rene for his help and suggestions during the research and refining this thesis. Further, acknowledgement to Dr. Eric Van Hullebusch, for his support during entire PhD program. I would like to thank my mentors Dr. Yu Zhang and Dr. Graciela Gonzalez-Gil, for their help with experimental designs and with manuscripts. I would like to acknowledge my thesis reviewers Dr. Cartsen Vogt and Dr. Artin Hatzikioseyian for their thorough revision and valualable feedback in this thesis.

I would like to acknowledge Dr. Jack Vossenberg for his help and advices in the microbial analysis. Further, acknowledgements to Dr. Filip Meysman and his team at the laboratory of Royal Netherlands Institute of Sea Research NIOZ, Yerseke, the Netherlands for providing the Lake Grevelingen sediments and fruitful discussions about the lake biochemistry. I would like to acknowledge Prof. Caroline P. Slomp and Dr. Matthias Egger for the collaborative work in the Lake Grevelingen.

I am grateful to my fellowship provider, the Erasmus Mundus Joint Doctorate Environmental Technologies for Contaminated Solids, Soils, and Sediments (ETeCoS^3) (FPA n° 2010-0009) which support the research and provide a special opportunity to study at UNESCO-IHE Institute for Water Education (The Netherlands), University of Cassino and Southern Lazio (Italy) and Shanghai Jiao Tong University (China). It was a great learning experience and multi-cultural combination to perform the research in three diverse country in three years.

Further, I would like to thank Zita Naangmenyele (MSc. student) for her contribution in this research work, especially in the chapter 3. Moreover, I would like to thank entire laboratory team of UNESCO-IHE for their support in the laboratory. Thanks to all the colleagues from the State Key Laboratory of Microbial Metabolism/Laboratory of Microbial Oceanography, Shanghai Jiao Tong University.

I would like to extend my acknowledgement to my colleague, Chiara Cassarini for all the collaborative works, moral support and friendship. It was great to walk along in the same path. Thanks to all the friends with whom I shared time during last three years in the Netherlands, Italy and China. Thanks to Carlos, Joy, Shrutika, Gabriele and Kirki for their help in different stages of this research. An especial thanks to my colleagues Dr. Jian Ding, Yan Wan Kai and Xiaoxia Liu for their support during my stay and work at China.

Finally, I would like to thanks my family, my parents who were always in my support. Thanks to my husband, Yubraj Gautam and my little one for all the love, support and always being by my side. The entire work would not be possible without them.

Summary

Anaerobic oxidation of methane (AOM) coupled to sulfate reduction (AOM-SR) is a biological process mediated by anaerobic methanotrophs (ANME) and sulfate reducing bacteria. Due to its relevance in regulating the global carbon cycle and potential biotechnological application for treating sulfate-rich wastewater, AOM-SR has drawn attention from the scientific community. However, the detailed knowledge on ANME community, its physiology and metabolic pathway are scarcely available, presumably due to the lack of either pure cultures or the difficulty to enrich the biomass. To enhance the recent knowledge on ANME distribution and enrichment conditions, this research investigated AOM-SR with the following objectives: (i) characterize the microbial communities responsible for AOM in marine sediment, (ii) enrich ANME in different bioreactor configurations, i.e. membrane bioreactor (MBR), biotrickling filter (BTF) and high pressure bioreactor (HPB), and (iii) assess the AOM-SR activity under different pressure and temperature conditions.

The microbes inhabiting coastal sediments from Marine Lake Grevelingen (the Netherlands) was characterized and the ability of the microorganisms to carry out AOM-SR was assessed. By performing batch activity tests for over 250 days, AOM-SR was evidenced by sulfide production and the concomitant consumption of sulfate and methane at approximately equimolar ratios and a sulfate reduction rate of 5 µmol g_{dw}^{-1} d^{-1} was attained. Sequencing of 16S rRNA genes showed the presence of ANME-3 in the Marine Lake Grevelingen sediment.

Two bioreactor configurations, i.e. MBR and BTF were operated under ambient conditions for 726 days and 380 days, respectively, to enrich the microorganisms from Ginsburg mud volcano performing AOM. The reactors were operated in fed-batch mode for the liquid phase with a continuous supply of gaseous methane. In the MBR, an external ultra-filtration membrane was used to retain the biomass, whereas, in the BTF, biomass retention was achieved via biomass attachment to the packing material. AOM-SR was recorded only after ~ 200 days in both bioreactor configurations. The BTF operation showed the enrichment of ANME in the biofilm by Illumina Miseq method, especially ANME-1 (40%) and ANME-2 (10%). Interestingly, in the MBR, aggregates of ANME-2 and *Desulfosarcina* were visualized by catalyzed reported deposition–fluorescence *in situ* hybridization (CARD-FISH). Acetate production was observed in the MBR, indicating that acetate was a possible intermediate of AOM. Although both bioreactor configurations showed good performance and resilience capacities for AOM enrichment, the sulfate reduction rate was slightly higher and faster in the BTF (1.3 mM d^{-1} on day 280) than the MBR (0.5 mM d^{-1} on day 380).

In order to simulate cold seep conditions and differentiate the impact of environmental conditions on AOM activities, sediment highly enriched with the ANME-2a clade was incubated in HPB at different temperature (4, 15 and 25°C at 10 MPa) and pressure (2, 10, 20 and 30 MPa at 15°C) conditions. The incubation at 10 MPa pressure and 15°C was observed to be the most suitable condition for the ANME-2a phylotype, which is similar to *in situ* conditions where the biomass was sampled, i.e. Captain Aryutinov mud volcano, Gulf of Cadiz. The incubations at 20 MPa and 30 MPa pressures showed the depletion in activities after 30 days of incubation. Incubation of AOM hosting sediment at *in situ* condition could be a preferred option for achieving high AOM activities and sulfate reduction rates.

In this thesis, it has been experimentally demonstrated that biomass retention and the continuous supply of methane can favor the growth of the slow growing anaerobic methane oxidizing community in bioreactors even under ambient conditions. Therefore, locating ANME habitats in shallow environments and enriching them at ambient conditions can be advantageous for future environmental biotechnology applications.

Samenvatting (Dutch)

Anaerobe oxidatie van methaan (AOM) gekoppeld aan sulfaat reductie (AOM-SR) is een biologisch proces dat wordt uitgevoerd door anaërobe methanotrofen (ANME) en sulfaat reducerende bacteriën. Vanwege de relevantie ervan in het reguleren de mondiale koolstofcyclus en potentiële biotechnologische toepassingen voor het verwijderen van metalen en sulfaat uit industriëel afvalwater, heeft AOM-SR de aandacht van de wetenschappelijke gemeenschap. Gedetailleerde kennis over de ANME, de fysiologie en metabole routes zijn schaars, waarschijnlijk als gevolg van enerzijds het ontbreken van reinkulturen of anderzijds de moeilijkheid om de biomassa op te hopen. Om de recente kennis over de ANME distributie en om hun ophoping te verbeteren, onderzocht dit onderzoek AOM-SR met de volgende doelstellingen: (i) kenmerken bepalen van de microbiële gemeenschappen die verantwoordelijk zijn voor AOM in mariene sedimenten, (ii) verrijken van ANME in verschillende bioreactor configuraties: membraan bioreactoren (MBR), biotrickling filters (BTF) en hoge druk bioreactoren (HPB), en (iii) bepalen van de AOM-SR activiteit onder verschillende druk en temperaturen.

De microorganismen die kustsedimenten van het Marine Grevelingen (Nederland) bewonen werden gekarakteriseerd en het vermogen van de micro-organismen om AOM-SR uit te voeren werd beoordeeld. Door het uitvoeren van batch activiteit testen van meer dan 250 dagen werd AOM-SR bevestigd via de sulfide productie en de daarmee gepaard gaande consumptie van sulfaat en methaan in equimolaire verhouding. Een sulfaat reductiesnelheid van 5 µmol g_{dw}^{-1} d^{-1} werd bereikt. Sequentieanalyse van de 16S rRNA genen toonde de aanwezigheid aan van ANME-3 in het Marine Grevelingenmeer sediment.

Twee bioreactorconfiguraties, d.w.z. MBR en BTF werden onder omgevingsomstandigheden gedurende, respectievelijk, 726 en 380 dagen bedreven om de AOM micro-organismen uit Ginsburg vulkaan sediment aan te rijken. De reactoren werden bedreven in fed-batch mode voor de vloeibare fase met een continue toevoer van gasvormig methaan. In de MBR werd een externe ultrafiltratie membraan gebruikt om de biomassa te behouden, terwijl biomassaretentie in de BTF werd bereikt via biomassa hechting aan het pakkingsmateriaal. AOM-SR kwam pas na ~ 200 dagen op gang. Verrijking van ANME in de BTF biofilm werd door Illumina

Miseq aangetoond en in het bijzonder ANME-1 (40%) en ANME-2 (10%) profileerden in de bioreactor. Interessant is dat in de MBR aggregaten van ANME-2 en *Desulfosarcina* zichtbaar werden gemaakt met behulp van CARD-FISH. Acetaat productie werd waargenomen in de MBR, wat aangeeft dat acetaat een mogelijk tussenproduct van AOM is. Hoewel beide bioreactor configuraties goede prestaties en een ophoping van AOM gaven, was de sulfaat reductie snelheid iets hoger in de BTF (1.3 mmol d^{-1} op dag 280) dan in de MBR (0.5 mmol d^{-1} op dag 380).

Om de omstandigheden zoals die zich voordoen in cold seeps te simuleren en de invloed van omgevingsfactoren op AOM activiteit te bepalen, werd een sediment aangerijkt met de ANME-2a clade geïncubeerd in HPB bij verschillende temperaturen (4, 15 en 25°C bij 10 MPa) en de druk (2, 10, 20 en 30 MPa bij 15°C). De beste omstandigheden zijn voor het ANME-2a phylotype waren de incubaties bij 10 MPa druk en 15°C, dit zijn omstandigheden vergelijkbaar met de *in situ* omstandigheden waarbij de biomassa werd bemonsterd, d.w.z. Capitein Aryutinov mud volcano, Golf van Cádiz. De incubaties bij 20 en 30 MPa toonden depletie in activiteit na 30 dagen incubatie. Incubatie van AOM bevattend sediment bij *in situ* condities is dus de voorkeursoptie voor het bereiken van hoge AOM activiteiten en sulfaat reductie snelheden.

In dit proefschrift is experimenteel aangetoond dat biomassaretentie en de continue toevoer van methaan de groei van de langzaam groeiende anaërobe methaan oxiderende gemeenschappen in bioreactoren kan bevorderen, zelfs onder omgevingsomstandigheden. Het lokaliseren van ondiepe ANME habitaten en ophoping van ANME bij omgevingsomstandigheden kunnen milieubiotechnologische toepassingen van AOM-SR een stap dichter bij brengen.

Sommario (Italian)

L'ossidazione anaerobica del metano (OAM) accoppiata alla solfato riduzione (OAM-SR) è un processo biologico mediato da metanotrofi anaerobici (ANME) e batteri solfato riduttori. Per via della sua importanza nel regolare il ciclo globale del carbonio e le sue potenziali applicazioni nel trattare le acque reflue ricche di solfati, l'OAM-SR ha ricevuto molte attenzioni da parte della comunità scientifica. Tuttavia, una conoscenza approfondita delle comunità ANME, delle sue fisiologie e delle sue vie metaboliche é ancora scarsa, presumibilmente a causa della mancanza di colture pure o della difficoltà nell'arricchire la biomassa. Per ampliare le conoscenze riguardo distribuzione e condizioni per l'arricchimento degli ANME, questo studio ha preso in esame l'OAM-SR con i seguenti obiettivi: (i) caratterizzare i microrganismi responsabili per l'OAM in sedimenti marini, (ii) arricchire gli ANME in differenti configurazioni reattoristiche, ovvero reattori biologici a membrana (RBM), letti percolatori (LP) e bioreattori ad alta pressione (BAP), e (iii) valutare l'OAM-SR a differenti condizioni di pressione e temperatura.

I microrganismi rinvenuti nei sedimenti costieri del lago salmastro Grevelingen (Paesi Bassi) sono stati caratterizzati, valutando le loro capacità nel portare a compimento l'OAM-SR. Conducendo test in condizioni batch sulle attività microbiche, per oltre 250 giorni, l'OAM-SR è stata dimostrata a partire dalla produzione di solfuro e dal concomitante consumo di solfato e metano in proporzioni approssimativamente equimolari, ottenendo un tasso di riduzione del solfato pari a 5 μmol g_{dw}^{-1} d^{-1} di massa secca per giorno. L'analisi sequenziale dei geni 16S rRNA ha mostrato la presenza di ANME-3 nei sedimenti del lago salmastro Grevelingen.

Due configurazioni reattoristiche, ovvero RBM e LP sono state operate in condizioni atmosferiche per 726 e 380 giorni, rispettivamente, per arricchire i microrganismi provenienti dal vulcano Ginsburg Mud, responsabili dell'OAM. I reattori sono stati operati in modalità fed-batch per quanto riguarda la fase liquida con supplementazione continua di metano gassoso. Nell'RBM, la biomassa è stata trattenuta tramite l'uso di una membrana ultra-filtrante esterna, mentre, nel LP, la biomassa era attaccata alla superficie del materiale all'interno del reattore. L'OAM-SR è stata notata solamente dopo 200 giorni circa in entrambe le configurazioni reattoristiche. Nel reattore LP, con il metodo Illumina Miseq è stato possibile rilevare l'arricchimento dell'ANME nel biofilm, in particolare ANME-1 (40%) e ANME-2 (10%). E' stato possibile visualizzare nel reattore RBM aggregati di ANME-2 e *Desulfosarcina* tramite tecnica CARD-FISH. Nel reattore RBM è stata inoltre osservata produzione di acetato, ad indicare che l'acetato è un plausibile intermedio dell'OAM. Sebbene entrambe le configurazioni reattoristiche abbiamo mostrato buone performance e una certa resilienza in relazione all'arricchimento dell'OAM, le cinetiche della solfato riduzione sono state leggermente più elevate nell'LP (1.3 mM giorno^{-1} al giorno 280) rispetto all'RBM (0.5 mM giorno^{-1} al giorno 380).

Del sedimento arricchito con il gruppo ANME-2a è stato incubato in BAP a differenti condizioni di temperature (4, 15 and 25°C at 10 MPa) e pressione (2, 10, 20 and 30 MPa a 15°C), per simulare le stesse condizioni di una sorgente marina fredda. Lo studio ha rivelato che l'incubazione a 10 MPa e 15°C è stata la condizione più idonea per il filotipo ANME-2; condizioni che sono infatti simili a quelle rilevate *in-situ*, dove la biomassa è stata campionata, ovvero il vulcano Captain Aryutinov Mud, nel golfo di Cadice. Le incubazioni a 20 e 30 MPa di pressione hanno mostrato un esaurimento delle attività dopo 30 giorni. L'incubazione del sedimento contenente microrganismi responsabili per l'OAM alle stesse condizioni di quelle rilevate *in situ* potrebbe rappresentare l'opzione da preferire per ottenere efficienze elevate di OAM e solfato riduzione.

In questa tesi, è stato dimostrato sperimentalmente che la conservazione della biomassa e la continua aggiunta di metano possono favorire la crescita dei microrganismi anaerobici metano-ossidanti in bioreattori, anche a condizioni atmosferiche. Di conseguenza, allocando gli habitat degli ANME in acque poco profonde e arricchendoli a condizioni atmosferiche può essere vantaggioso per future applicazioni nel campo delle biotecnologie ambientali.

Résumé (French)

L'oxydation anaérobie du méthane (AOM) couplé à la réduction du sulfate (AOM-SR) est un processus biologique médié par méthanotrophes anaérobie (ANME) et de bactéries sulfato-réductrices. La communauté scientifique s'inquiète de AOM, en raison de sa pertinence dans la régulation du cycle global du carbone et de la potentielle application biotechnologique pour le traitement de sulfate riches eaux usées. Cependant, la connaissance détaillée des micro-organismes responsables de ce bioprocessus, de la physiologie et de la voie métabolique sont rarement disponibles dans la littérature, probablement en raison de l'absence des cultures pures ou des difficultés d'enrichir la biomasse. Pour améliorer les connaissances récentes sur les conditions de distribution et d'enrichissement ANME, cette recherche a étudié AOM-SR avec les objectifs suivants: (i) caractériser les communautés microbiennes responsables de AOM dans les sédiments marins, (ii) de les enrichir dans les bioréacteurs avec différentes configurations, à savoir bioréacteur à membrane (MBR), filtre biotrickling (BTF) et bioréacteur à haute pression (HPB), et (iii) d'évaluer l'activité de l'ANME et le processus AOM dans différentes conditions de pression et de température.

Les microbes habitant la zone de transition du sulfate-méthane peu profonde dans les sédiments de Marine lac Grevelingen (Pays-Bas) ont été caractérisés et leur capacité de faire AOM-SR a été évaluée. Un test d'activité a été réalisée en discontinu pour 250 jours, AOM-SR est mise en évidence par la production de sulfure et de la prise concomitante de sulfate et de méthane dans des rapports équimolaires et il a été atteint 5 μmol g_{dw}^{-1} d^{-1} de taux de réduction du sulfate. L'analyse des séquences de gènes rRNA 16S a montré la présence de méthanotrophes anaérobie ANME-3 dans les sédiments marins du lac Grevelingen.

Deux configurations de bioréacteurs, à savoir MBR et BTF ont été opérés dans des conditions ambiantes pendant 726 jours et 380 jours, respectivement, pour enrichir les micro-organismes de Ginsburg Mud Volcano performantes AOM. Les réacteurs sont exploités en mode fed-batch pour la phase liquide avec un apport continu de méthane. Dans le MBR, une membrane d'ultrafiltration externe a été utilisée pour retenir la biomasse, alors que, dans la BTF, la rétention de biomasse a été accomplie par la fixation de la biomasse sur le matériau d'emballage. AOM-SR a été enregistrée seulement après ~ 200 jours dans les deux configurations de bioréacteurs. L'opération du BTF a montré l'enrichissement de l'ANME dans le biofilm par la méthode Illumina Miseq, en particulier ANME-1 (40%) et ANME-2 (10%). Dans le MBR, les agrégats d'ANME-2 et *Desulfosarcina* ont été visualisées par CARD-FISH. La production d'acétate a été observée dans le MBR, ce qui indique que l'acétate était un possible intermédiaire d'AOM. Bien que les deux configurations de bioréacteurs ont montré de bonnes performances, le taux de réduction du sulfate était légèrement plus élevée et plus rapide dans la BTF (1.3 mM par jour âpres 280 jours) que le MBR (0.5 mM par jour jour âpres 380 jours).

Afin de simuler les conditions de suintement froid et de différencier l'impact des conditions environnementales sur AOM, les sédiments fortement enrichi avec le clade ANME-2a ont été incubées dans HPB à différentes températures (4, 15 et 25°C à 10 MPa) et pressions (2, 10, 20 et 30 MPà 15°C). L'incubation à une pression de 10 MPa et 15°C a été observé comme la condition la plus appropriée pour la phylotype ANME-2a, quiest similaire aux conditions in situ (Capitaine Aryutinov Mud Volcano, Golfe de Cadix). Les incubations à la pression de 20 et 30 MPa ont montré la diminution des activités après 30 jours d'incubation. L'incubation de ce sédiment aux conditions *in situ* pourrait être une option privilégiée pour obtenir une activité AOM-SR plus élevée.

Dans cette thèse, il a été démontré expérimentalement que la rétention de la biomasse et l'approvisionnement continu de méthane peuvent favoriser la croissance de la lente communauté microbienne qui oxyde le méthane en anaérobiose dans des bioréacteurs, même dans des conditions ambiantes. Par conséquent, la localisation des habitats de ANME dans des environnements peu profonds et l'enrichissant dans des conditions ambiantes peut être avantageuse pour les futures applications de la biotechnologie environnementale.

CHAPTER 1

General Introduction and Thesis Outline

1.1 General introduction and problem statement

In marine environments, which cover almost 70% of the earth's total surface, the methane cycle is exclusively controlled by microbial activities with an estimated annual rate of methanogenesis of 85-300 Tg methane year^{-1} (Conrad, 2009). Out of the total amount of methane produced in sea sediments, ca. 90% is consumed by microbially mediated anaerobic oxidation of methane (AOM) (Hinrichs & Boetius, 2002; Reeburgh, 2007). Geochemical evidences, isotopic data, and microbial process measurements indicate that AOM is coupled with the reduction of sulfate in marine sediments which is represented by Eq. 1.1 (Nauhaus et al., 2002):

$$CH_4 + SO_4^{2-} \rightarrow HCO_3^- + HS^- + H_2O \qquad \Delta G° = -17 \text{ kJ mol}^{-1} \quad \text{(Eq. 1.1)}$$

The AOM coupled to sulfate reduction (AOM-SR) process involves methane oxidation by anaerobic methanotrophic archaea (ANME), while sulfate in the sub-surface of the marine sediment is reduced by sulfate reducing bacteria (SRB). These organisms oxidize methane to carbon dioxide while reducing sulfate to sulfide (Eq. 1.1), splitting the energy supplied by this coupled process between them. So far, three types of ANME clades have been identified for AOM which depend on sulfate as the terminal electron acceptor (ANME-1, ANME-2, ANME-3) (Boetius et al., 2000; Hinrichs et al., 1999; Knittel et al., 2005; Niemann et al., 2006; Bhattarai et al., 2017). All ANME are phylogenetically related to various groups of methanogenic Archaea. ANME-2 and ANME-3 are clustered within the order *Methanosarcinales*, while ANME-1 belongs to a new order which is distantly related to the orders *Methanosarcinales* and *Methanomicrobiales*. Moreover, ANME-3 is closely related to the genus *Methanococcoides* of methanogens. The habitat for the ANME is widely distributed globally along the marine locations such as cold methane seeps (Orcutt et al., 2005), mud volcanoes (Niemann et al., 2006), hydrothermal vents (Teske et al., 2002), coastal sediment (Treude et al., 2005; Oni et al., 2015; Bhattarai et al., 2017) and hypersaline environments (Lloyd et al., 2006).

Different theories have been postulated for the electron transfer mechanisms between ANME and SRB; however, the explicit knowledge in physiology and mechanism is yet limited. Milucka et al. (2012) revealed a distinct mechanism for AOM with the possible involvement of ANME only for AOM and elemental sulfur being an intermediate for sulfur metabolism. Other recent studies pointed towards cooperative interactions between ANME and SRB based on interspecies electron transfer, both for thermophilic ANME-1 and mesophilic ANME-2 consortia (McGlynn et al., 2015; Wagner, 2015). However, Scheller et al. (2016) showed that ANME-2 and SRB can act singly for AOM by using different electron acceptors, such as ferric iron.

AOM-SR has mainly two relevances, one is ecological, i.e., prevention of methane emission to the atmosphere and another potential biotechnological applications. AOM is restricting a large amount of methane to the sea floor with a significant effect in global warming (Boetius & Wenzhofer, 2013; Reeburgh, 2007). Moreover, AOM-SR is a biological process with potential application for the desulfurization of wastewater, where methane can be used as a sole electron donor. Biological desulfurization under anaerobic conditions is a well known (bio)technique for the removal of sulfur compound and metals from wastewater stream and precipitates as metal sulfide (Lens et al., 2002). Electron donors are essential for the treatment of sulfate-containing wastewaters by biological methods as energy source for the microorganisms involved in the desulfurization process. However, most industrial wastewaters are deficient in dissolved organic carbon; hence, supply of an external carbon source is essential (Reyes-Alvarado et al., 2017). One of the challenging factors for wastewater treatment industry is the cost of electron donors, which can be overcome by utilizing the cheapest electron donor methane via the AOM process (Gonzalez-Gil et al., 2011).

Application of AOM-SR has several benefits over other common electron donors (hydrogen, ethanol, acetate): cheaper electron donor, less conversion energy is required, minimal substrate loss by avoiding unwanted acetogenesis/methanogenesis and no need of carbon monoxide removal (Meulepas et al., 2010). Thus, it is appealing to optimize the sulfate reduction process with methane. However, the very slow growing nature of ANME, with least doubling time of 1.5-7.0 months and low solubility of methane at ambient pressure (1.44 mM in distilled water at 20°C) is highly challenging for the operation of bioreactors. For tackling these challenges, different biomass retention approaches are applied, i.e. membrane bioreactors (Meulepas et al., 2009), high pressure reactors for high methane solubility (Zhang et al., 2011) and provision of biomass attachment sponges

(Aoki et al., 2014; Cassarini et al., 2017). Therefore, the AOM-SR study in different reactor configurations is one of the focuses of this research (Figure 1.1). Other important aspect is to have the ample knowledge of ANME community and mechanisms, for example, either they form aggregates or remain as single cells, insight into the sulfate reducing partners and effects of different environmental conditions to the AOM-SR activity.

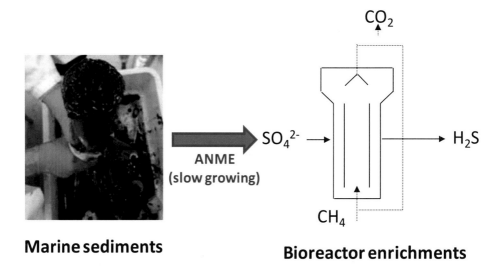

Marine sediments   **Bioreactor enrichments**

Figure 1.1 Illustration depicting the rationale of this PhD study, which represent the collection of (deep or shallow sea) sediment hosting AOM and the enrichment of microbial community in bioreactors.

1.2 Objectives and scope of the study

The main objective of this research is to enrich the AOM performing microbial community in bioreactors and assess its performance along with microbial community characterization.

The specific objectives are:

1. To investigate the physiology of AOM and identify the factors controlling the distribution of ANME

2. To assess the sulfate reducing activity of sediment from Marine Lake Grevelingen by using alternative electron acceptors and electron donors

3. To characterize the AOM hosting anaerobic marine sediment from Marine Lake Grevelingen by the activity measurements and microbial community analysis

4. To enrich the AOM community from the Gulf of Cadiz in a biotrickling filter bioreactor and study their activities and change in microbial community composition during its operation To study AOM activity by Gulf of Cadiz sediment in a membrane bioreactor

5. To assess the response of a highly enriched ANME-2a community to different pressure and temperature gradients in high pressure bioreactors

1.3 Outlines of this thesis

This dissertation is divided into eight chapters. The first chapter (Chapter 1) provides a brief overview of this PhD research and the thesis as depicted by Figure 1.2.

In chapter 2, the current situation of ANME studies, recent findings, development of new study tools and its constraints will be discussed. One of the bottlenecks for the AOM studies is the less known physiology and mechanism of ANME, which will be detailed in this chapter. Moreover the distribution of ANME and the environmental factors responsible for its distribution will be discussed.

Chapters 3 and 4 focus on the characterization of the marine sediment originating from Marine Lake Grevelingen (the Netherlands), which host diverse microbial communities with the potential role in carbon and sulfur cycle. The sulfate reduction rate comparison among different electron donors and potential AOM with different sulfur compounds by the sediment from Marine Lake Grevelingen will be presented in chapter 3. In chapter 4, the AOM activity of the sediment will be assessed and the microbial community responsible for AOM will be detailed.

Chapters 5 and 6 deal with the enrichment and activity study of AOM community in the bioreactors which are operated at ambient conditions. Chapter 5 will mainly focus on the ANME enrichment in specially designed biotrickling filter type bioreactor and demonstrate the AOM-SR activities. Chapter 6 will be the synthesis of the findings from a long term AOM activity study in a continuously operated membrane bioreactor (at ambient pressure) and activity measurements during the membrane bioreactor run.

Chapter 7 will illustrate the findings on the ANME physiology by studying the effect of two environmental factors, i.e. pressure and temperature, on the activity of a highly enriched ANME-2a community. This chapter will explore the potential changes on the AOM activity and involved microbial community with different temperature and pressure gradients.

Chapter 8 synthesizes the findings of all chapters of this thesis and will provide a general discussion on this PhD research. Chapter 8 thus provides an overview on the implication of the findings of this research and suggest ways forward for future research and applications.

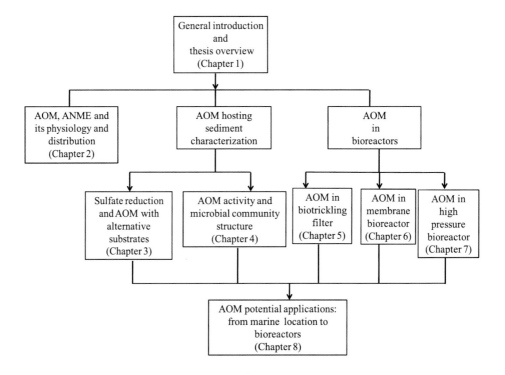

Figure 1.2 *Overview of the structure of this PhD thesis.*

1.4 References

Aoki, M., Ehara, M., Saito, Y., Yoshioka, H., Miyazaki, M., Saito, Y., Miyashita, A., Kawakami, S., Yamaguchi, T., Ohashi, A., Nunoura, T., Takai, K., Imachi, H. 2014. A long-term cultivation of an anaerobic methane-oxidizing microbial community from deep-sea methane-seep sediment using a continuous-flow bioreactor. *PLoS ONE*, **9**(8), 1-14.

Bhattarai, S., Cassarini, C., Gonzalez-Gil, G., Egger, M., Slomp, C.P., Zhang, Y., Esposito, G., and Lens, P.N. 2017. Anaerobic methane-oxidizing microbial community in a coastal marine sediment: anaerobic methanotrophy dominated by ANME-3. *Microb. Ecol.*, **74**(3), 608-622.

Boetius, A., Ravenschlag, K., Schubert, C.J., Rickert, D., Widdel, F., Gieseke, A., Amann, R., Jorgensen, B.B., Witte, U., Pfannkuche, O. 2000. A marine microbial consortium apparently mediating anaerobic oxidation of methane. *Nature*, **407**(6804), 623-626.

Boetius, A., Wenzhofer, F. 2013. Seafloor oxygen consumption fuelled by methane from cold seeps. *Nat. Geosci.*, **6**(9), 725-734.

Cassarini, C., Rene, E. R., Bhattarai, S., Esposito, G., Lens, P. N. 2017. Anaerobic oxidation of methane coupled to thiosulfate reduction in a biotrickling filter. *Biores. Technol.*, **240**, 214-222.

Conrad, R. 2009. The global methane cycle: recent advances in understanding the microbial processes involved. *Environ. Microbiol. Rep.*, **1**(5), 285-292.

Gonzalez-Gil, G., Meulepas, R.J.W., Lens, P.N.L. 2011. Biotechnological aspects of the use of methane as electron donor for sulfate reduction. in: Murray, M.-Y. (Ed.),*Comprehensive Biotechnology*. Vol. 6. (2nd edition), Elsevier B.V., Amsterdam, the Netherlands, pp. 419-434.

Hinrichs, K.-U., Boetius, A. 2002. The Anaerobic Oxidation of Methane: New Insights in Microbial Ecology and Biogeochemistry. in: Wefer, G., Billett, D., Hebbeln, D., Jørgensen, B., Schlüter, M., van Weering, T.E. (Eds.), *Ocean Margin Systems*. Springer Berlin Heidelberg, Germany, pp. 457-477.

Hinrichs, K.U., Hayes, J.M., Sylva, S.P., Brewer, P.G., DeLong, E.F. 1999. Methane-consuming archaebacteria in marine sediments. *Nature*, **398**(6730), 802-805.

Knittel, K., Lösekann, T., Boetius, A., Kort, R., Amann, R. 2005. Diversity and distribution of methanotrophic archaea at cold seeps. *Appl. Environ. Microbiol.*, **71**(1), 467-479.

Lens, P., Vallerol, M., Esposito, G., Zandvoort, M. 2002. Perspectives of sulfate reducing bioreactors in environmental biotechnology. *Rev. Environ. Sci. Biotechnol.*, **1**(4), 311-325.

Lloyd, K.G., Lapham, L., Teske, A. 2006. An anaerobic methane-oxidizing community of ANME-1b archaea in hypersaline Gulf of Mexico sediments. *Appl. Environ. Microbiol.*, **72**(11), 7218-7230.

McGlynn, S.E., Chadwick, G.L., Kempes, C.P., Orphan, V.J. 2015. Single cell activity reveals direct electron transfer in methanotrophic consortia. *Nature*, **526**, 531-535.

Meulepas, R., Stams, A., Lens, P. 2010. Biotechnological aspects of sulfate reduction with methane as electron donor. *Rev. Environ. Sci. Biotechnol.*, **9**(1), 59-78.

Meulepas, R.J.W., Jagersma, C.G., Gieteling, J., Buisman, C.J.N., Stams, A.J.M., Lens, P.N.L. 2009. Enrichment of anaerobic methanotrophs in sulfate-reducing membrane bioreactors. *Biotechnol. Bioeng.*, **104**(3), 458-470.

Milucka, J., Ferdelman, T.G., Polerecky, L., Franzke, D., Wegener, G., Schmid, M., Lieberwirth, I., Wagner, M., Widdel, F., Kuypers, M.M. 2012. Zero-valent sulphur is a key intermediate in marine methane oxidation. *Nature*, **491**(7425), 541-546.

Nauhaus, K., Boetius, A., Krüger, M., Widdel, F. 2002. In vitro demonstration of anaerobic oxidation of methane coupled to sulphate reduction in sediment from a marine gas hydrate area. *Environ. Microbiol.*, **4**(5), 296-305.

Niemann, H., Losekann, T., de Beer, D., Elvert, M., Nadalig, T., Knittel, K., Amann, R., Sauter, E.J., Schluter, M., Klages, M., Foucher, J.P., Boetius, A. 2006. Novel microbial communities of the Haakon Mosby mud volcano and their role as a methane sink. *Nature*, **443**(7113), 854-858.

Oni, O. E., Miyatake, T., Kasten, S., Richter-Heitmann, T., Fischer, D., Wagenknecht, L., Ksenofontov, V., Kulkarni, A., Blumers, M., Shylin, S. 2015. Distinct microbial populations are tightly linked to the profile of dissolved iron in the methanic sediments of the Helgoland mud area, North Sea. *Front. Microbiol.* **6**(365), 1-15.

Orcutt, B., Boetius, A., Elvert, M., Samarkin, V., Joye, S.B. 2005. Molecular biogeochemistry of sulfate reduction, methanogenesis and the anaerobic oxidation of methane at Gulf of Mexico cold seeps. *Geochim. Cosmochim. Acta.*, **69**(17), 4267-4281.

Reeburgh, W.S. 2007. Oceanic methane biogeochemistry. *Chem. Rev.*, **107**(2), 486-513.

Reyes-Alvarado, L. C., Camarillo-Gamboa, Á., Rustrian, E., Rene, E. R., Esposito, G., Lens, P. N., Houbron, E. 2017. Lignocellulosic biowastes as carrier material and slow release electron donor for sulphidogenesis of wastewater in an inverse fluidized bed bioreactor. *Environ. Sci. Pollut. Res.*, 1-14.

Scheller, S., Yu, H., Chadwick, G.L., McGlynn, S.E., Orphan, V.J. 2016. Artificial electron acceptors decouple archaeal methane oxidation from sulfate reduction. *Science*, **351**, 703-707.

Teske, A., Hinrichs, K.U., Edgcomb, V., Gomez, A.D., Kysela, D., Sylva, S.P., Sogin, M.L., Jannasch, H.W. 2002. Microbial diversity of hydrothermal sediments in the Guaymas Basin: Evidence for anaerobic methanotrophic communities. *Appl. Environ. Microbiol.*, **68**(4), 1994-2007.

Treude, T., Krüger, M., Boetius, A., Jørgensen, B. B. 2005. Environmental control on anaerobic oxidation of methane in the gassy sediments of Eckernförde Bay (German Baltic). *Limnol. Oceanogr.* **50**(6), 1771-1786.

Wagner, M. 2015. Microbiology: Conductive consortia. *Nature*, **526**, 513-514.

Zhang, Y., Maignien, L., Zhao, X., Wang, F., Boon, N. 2011. Enrichment of a microbial community performing anaerobic oxidation of methane in a continuous high-pressure bioreactor. *BMC Microbiol.*, **11**(1), 1-8.

CHAPTER 2

Physiology and Distribution of Anaerobic Oxidation of Methane by Archaeal Methanotrophs

Abstract

Methane is oxidized in marine anaerobic environments, where sulfate rich sea water meets biogenic or thermogenic methane. In those niches, few phylogenetically distinct microbial types, i.e. anaerobic methanotrophs (ANME), are able to grow through anaerobic oxidation of methane (AOM). Due to the relevance of methane in the global carbon cycle, ANME draw the attention of a broad scientific community since five decades. This chapter presents and discusses the microbiology and physiology of ANME up to the recent discoveries, revealing novel physiological types of anaerobic methane oxidizers which challenge the view of obligate syntrophy for AOM. The drivers shaping the distribution of ANME in different marine habitats, from cold seep sediments to hydrothermal vents, are overviewed. Multivariate analyses of the abundance of ANME in various habitats identify a distribution of distinct ANME types driven by the mode of methane transport. Intriguingly, ANME have not yet been cultivated in pure culture, despite of intense attempts. Further, advances in understanding this microbial process are hampered by insufficient amounts of enriched cultures. This review discusses the advantages, limitations and potential improvements for ANME cultivation systems and AOM study approaches.

2.1 Introduction

Methane (CH_4) is the most abundant and completely reduced form of hydrocarbon. It is the most stable hydrocarbon, which demands +439 kJ mol^{-1} energy to dissociate the hydrocarbon bond (Thauer & Shima, 2008). CH_4 is a widely used energy source, but it is also the second largest contributor to human induced global warming, after carbon dioxide (CO_2). CH_4 concentrations in the atmosphere have increased from about 0.7 to 1.8 ppmv (i.e. an increase of 150%) in last 200 years, and experts estimate that this increase is responsible for approximately 20% of the Earth's warming since pre-industrial times (Kirschke et al., 2013). On a per mol basis and over a 100 year horizon, the global warming potential of CH_4 is about 25 times more than that of CO_2 (IPCC, 2007). Therefore, large scientific efforts are being made to resolve detailed maps of CH_4 sources and sinks, and how these are affected by the increased levels of this gas in the atmosphere (Kirschke et al., 2013).

12

The global CH_4 cycle is largely driven by microbial processes of CH_4 production (i.e. methanogenesis) and CH_4 oxidation (i.e. methanotrophy). CH_4 is microbially produced by the anaerobic degradation of organic compounds or through CO_2 bioreduction (Nazaries et al., 2013). These CH_4 production processes occur in diverse anoxic subsurface environments like rice paddies, wetlands, landfills, contaminated aquifers as well as freshwater and ocean sediments (Reeburgh, 2007). CH_4 can also be formed physio-chemically at specific temperatures of about 150°C to 220°C (thermogenesis). It is estimated that more than half of the CH_4 produced globally is oxidized microbially to CO_2 before it reaches the atmosphere (Reeburgh, 2007). Both aerobic and anaerobic methanotrophy are the responsible processes. The first involves the oxidation of CH_4 to methanol in the presence of molecular oxygen (and subsequently to CO_2) by methanotrophic bacteria (Chistoserdova et al., 2005; Hanson & Hanson, 1996), whereas the second includes the oxidation of CH_4 to CO_2 in the absence of oxygen by a clade of archaea, called anaerobic methanotrophs (ANME) and the process is known as the anaerobic oxidation of methane (AOM).

Large CH_4 reservoirs on Earth, i.e.450 to 10,000 Gt carbon (Archer et al., 2009; Wallmann et al., 2012) are found as CH_4 hydrates beneath marine sediments, mostly formed by biogenic processes (Pinero et al., 2013). CH_4 hydrates, or CH_4 clathrates, are crystalline solids, consisting of large amounts of CH_4 trapped by interlocking water molecules (ice). They are stable at high pressure (> 60 bar) and low temperature (< 4°C) (Boetius & Wenzhöfer, 2013; Buffett & Archer, 2004), and are typically found along continental margins at depths of 600 to 3000 m below sea level (Archer et al., 2009; Boetius & Wenzhöfer, 2013; Reeburgh, 2007). By gravitational and tectonic forces, CH_4 stored in hydrate seeps into the ocean sediment under the form of mud volcanoes, gas chimneys, hydrate mounds and pock marks (Boetius & Wenzhöfer, 2013). These CH_4 seepage manifestations are environments where AOM has been documented (Table 2.1) e.g. Black Sea carbonate chimney (Treude et al., 2007), Gulf of Cadiz mud volcanoes (Niemann et al., 2006a), Gulf of Mexico gas hydrates (Joye et al., 2004). Besides, AOM also occurs in the sulfate-methane transition zones (SMTZ) of coastal sediments (Treude et al., 2005b; Oni et al., 2015; Bhattarai et al., 2017). The SMTZ are quiescent sediment environments, where the upwards diffusing (thermogenic and biogenic) CH_4 is oxidized when it meets sulfate

(SO_4^{2-}), which is transported downwards from the overlaying seawater (upper panel of Figure 2.1). Considering that SO_4^{2-} is abundant in seawater and that oxygen in sea bed sediments is almost absent, AOM coupled to the reduction of SO_4^{2-} is likely the dominant biological sink of CH_4 in these environments.

It is estimated that CH_4 seeps, which are generally lying above CH_4 hydrates (Suess, 2014), annually emit 0.01 to 0.05 Gt C, contributing to 1 to 5% of the global CH_4 emissions to the atmosphere (Boetius & Wenzhöfer, 2013). These emissions would be higher if CH_4 was not scavenged by aerobic or anaerobic oxidation of CH_4. While aerobic CH_4 oxidation is dominant in shallow oxic seawaters (Tavormina et al., 2010), AOM is found in the anoxic zones of the sea floor (Knittel & Boetius, 2009; Reeburgh, 2007; Wankel et al., 2010). Due to limited data, it has not been possible to determine the exact global values of CH_4 consumption by AOM. But, the AOM in the SMTZ and CH_4 seep environments has been tentatively estimated at 0.05 Gt C and 0.01 Gt C per year, respectively (Boetius & Wenzhöfer, 2013).

Besides the biogeochemical implications of AOM, this microbial process can have biotechnological applications for the treatment of waste streams rich in SO_4^{2-} or nitrate/nitrite but low in electron donor. Recently, few studies have highlighted the prospective of AOM and ANME in environmental biotechnology, where CH_4 is used as the sole electron donor to achieve SO_4^{2-} reduction (SR) in bioreactors (Gonzalez-Gil et al., 2011; Meulepas et al., 2009a; Meulepas et al., 2010c). Biological SR is a well-known technique to remove sulfur and metals from wastewaters. Following the removal of sulfur, metals can be recovered by metal sulfide precipitation (Lens et al., 2002). Many industrial wastewaters are deficient in dissolved organic carbon. Hence, supplementation of external carbon sources and electron donors is essential for microbial SR. Frequently used electron donors for SO_4^{2-} reducing treatment plants are hydrogen/CO_2 and ethanol (Widdel & Hansen, 1992), which are costly and can be replaced by low-priced electron donors (Gonzalez-Gil et al., 2011). It is estimated that the overall treatment costs would be reduced by a factor of 2 to 4 if CH_4 from natural gas or biogas would be used in SO_4^{2-} reducing bioreactors as an electron donor instead of hydrogen or ethanol (Meulepas et al., 2010c). The major limitation identified for the biotechnological application of AOM is the extremely slow growth rates of the ANME, currently with doubling times as high as 2-7 months (Meulepas et al., 2010c).

14

A recent innovative idea is the use of key AOM enzymes for the biotechnological conversion of CH_4 to liquid fuels at high carbon conservation efficiencies (Haynes & Gonzalez, 2014). CH_4 could be transformed into butanol efficiently, if enzymes responsible for AOM activate CH_4 and assist in C-C bond formation (Haynes & Gonzalez, 2014). This concept is of interest because logistics and infrastructure for handling liquid fuels are more cost effective than those for utilizing compressed natural gas. A detailed elucidation of the ANME metabolism is a prerequisite to the development of such biotechnological applications of AOM.

2.2 Microbiology of anaerobic methane oxidation

2.2.1 Discovery of AOM

AOM coupled to SO_4^{2-} reduction (AOM-SR) takes place where SO_4^{2-} meets either biogenic or thermogenic CH_4. This unique microbiological phenomenon, AOM, was recognized since five decades as a key to close the balance of oceanic carbon (Martens & Berner, 1974; Reeburgh, 1976). Since then, various key discoveries have elucidated the AOM process to some extent, but its exact biochemical mechanism is still unclear (Figure 2.1). The AOM was first deduced from CH_4 and SO_4^{2-} profile measurements in marine sediments (Iversen & Jorgensen, 1985; Reeburgh, 1980; Zehnder & Brock, 1980). Occurrence of AOM yields typical concave-up CH_4 profiles in sediment columns with high CH_4 concentrations in the deep sediment layers and very low CH_4 concentrations at the sediment water interface (Figure 2.1).

Quasi *in situ* and *in vitro* studies using radiotracers confirmed AOM as a biological process but the microbes mediating AOM was not defined (Iversen & Jorgensen, 1985; Reeburgh, 1980; Zehnder & Brock, 1980). Additional *in vitro* studies suggested that the AOM process was performed by a unique microbial community (Boetius et al., 2000; Hinrichs et al., 1999; Hoehler et al., 1994): the anaerobic methanotrophs (ANME), mostly in association with SO_4^{2-} reducing bacteria (SRB) (Figure 2.2). The identification of the microorganisms involved in AOM is crucial to explain how CH_4 can be efficiently oxidized with such a low energy yield ($\Delta G^\circ = -17$ KJ mol^{-1} CH_4).

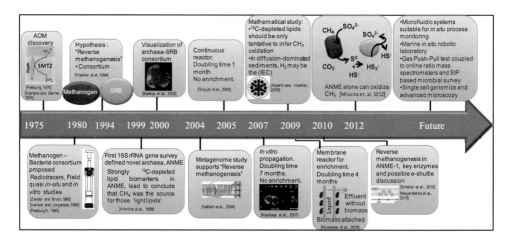

Figure 2.1 *Time line of relevant research and discoveries on the AOM-SR. The major milestones achieved are depicted in their respective year along with some future possibility in the AOM studies.*

By fluorescence *in situ* hybridization (FISH) based visualizations with specifically designed probes, the *in situ* occurrence of such archaea-bacteria associations was recorded, showing that the ANME-groups are widely distributed throughout marine sediments (Bhattarai et al., 2017; Boetius et al., 2000; Knittel et al., 2005; Orphan et al., 2002; Schreiber et al., 2010).

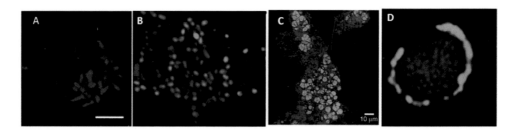

Figure 2.2 *Fluorescence in situ hybridization images from different ANME types. A) Single ANME-1 in elongated rectangular shape (red color) inhabiting as mono specific clade in the Guaymas Basin hydrothermal vent (Holler et al., 2011b), B) Aggregate of cocci shaped ANME-2 (red color) and DSS (green color), enrichment sample after 8 years from the Isis mud volcano in the Mediterranean Sea. The image was taken from the web: http://www.mpg.de/6619070/marine-methane-oxidation, C) Aggregate of large densely clustered ANME-2d (green) and other bacteria (blue color) obtained from a bioreactor enrichment (Haroon et al., 2013) and D) Aggregate of cocci shaped ANME-3 (red color) and DBB (green color) inhabiting Haakon Mosby mud volcano (Niemann et al., 2006b).*

The physico-chemical drivers shaping the global distribution of ANME consortia are not fully resolved to date (see section: Drivers for distribution of ANME). Instead,

AOM activity tests and *in vitro* studies allowed the estimation of their doubling time in the order of 2-7 months, realizing the extremely slow growth of ANME (Nauhaus et al., 2007).

2.2.2 ANME phylogeny

Based on the phylogenetic analysis of 16S rRNA genes (Figure 2.3A), ANME have been grouped into three distinct clades, i.e. ANME-1, ANME-2 and ANME-3 (Bhattarai et al., 2017; Boetius et al., 2000; Hinrichs et al., 1999; Knittel et al., 2005; Niemann et al., 2006b). All ANME are phylogenetically related to various groups of methanogenic archaea (Figure 2.3). ANME-2 and ANME-3 are clustered within the order *Methanosarcinales*, while ANME-1 belongs to a new order which is distantly related to the orders *Methanosarcinales* and *Methanomicrobiales* (Figure 2.3). Specifically, ANME-3 is closely related to the genus *Methanococcoides*. FISH analysis showed that microorganisms belonging to the ANME-2 and ANME-3 groups are cocci-shaped, similar to *Methanosarcina* and *Methanococus* methanogens (Figure 2.2B and 2.2D). On the contrary to ANME-2 and ANME-3, ANME-1 mostly exhibits a rod-shape morphology (Figure 2.2A).

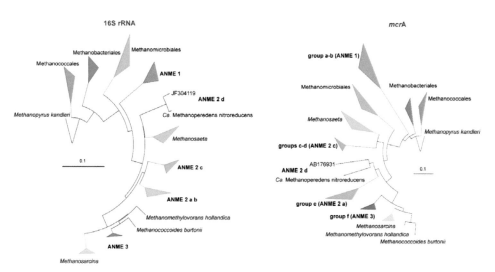

Figure 2.3 Phylogenetic affiliation of anaerobic methanotrophs (ANME) based on the 16S rRNA and mcrA genes. The 16S rRNA and mcrA sequences, retrieved from NCBI databases, were respectively aligned with the SINA aligner (Pruesse et al., 2012) and the Clustal method as previously described (Hallam et al., 2003). Both trees were inferred using the Neighbor-Joining method (Saitou & Nei, 1987). Bars refer to 10% estimated distance.

Samples from marine habitats around the globe have been retrieved through the years, and extensive molecular analysis of such samples have yielded a great number of 16S rRNA and *mcr*A gene sequences of the archaeal microorganisms inhabiting those sites. The *mcr*A gene encodes the alpha-subunit of the methyl coenzyme-M reductase (MCR), which is catalyzing the last step in methanogenesis, and it is also thought to catalyze the oxidation of CH_4, since MCR can function in reverse methanogenesis (Kruger et al., 2003; Scheller et al., 2010). Upon phylogenetic analysis based on 16S rRNA (Figure 2.3A) and *mcr*A (Figure 2.3B) genes, the three major groups of ANME were identified. ANME-1 is further subgrouped into ANME-1a and ANME-1b. ANME-2 is divided into four subgroups i.e. ANME-2a, ANME-2b, ANME-2c and ANME-2d, whereas, so far no subgroups of ANME-3 have been defined (Figure 2.3). The *mcr*A genes phylogeny of the various archaeal orders closely parallels that of the 16S rRNA genes (Figure 2.3).

The crystal structure of the MCR isolated from a Black Sea mat, naturally enriched in ANME-1, reveals that ANME-1 MCR and methanogenic MCR have distinct features in and around the active site of the enzyme (Shima et al., 2012). In the MCR from ANME-1, the prosthetic group is not the conventional F_{430} cofactor but a methylthio-F_{430} variant. To accommodate the variant of the cofactor, the geometry of the active site of the ANME-1 MCR enzyme is modified so that the amino acid glutamine, which post-translationally is 2-C methylated and thus bulky, is replaced by a small valine molecule (Mayr et al., 2008; Shima et al., 2012). These distinct features might reflect that ANME-1 are more distantly affiliated to the other ANME groups, for which no such modifications of their MCR seem to exist (Shima et al., 2012).

Besides their close phylogenetic relationships, ANME exhibit other similarities with methanogenic archaea. For example, sequenced genomes of ANME-1 and ANME-2 from environmental samples indicate that, except for the N^5, N^{10}-methylene-tetrahydromethanopterin (H$_4$MPT) reductase in the ANME-1 metagenome (Meyerdierks et al., 2010), these ANME contain homologous genes for the enzymes involved in all the seven steps of methanogenesis from CO_2 (Haroon et al., 2013; Meyerdierks et al., 2010; Wang et al., 2014). Furthermore, with the exception of coenzyme M-S-S-coenzyme B heterodisulfide reductase, all those enzymes catalyzing the CH_4 formation were confirmed to catalyze reversible reactions (Thauer, 2011; Scheller et al., 2010). Thus, it is hypothesized that ANME oxidize CH_4

via methanogenic enzymatic machinery functioning in reverse, i.e., reversal of CO_2 reduction to CH_4 (Hallam et al., 2004; Meyerdierks et al., 2010).

2.2.3 AOM coupled to sulfate reduction

The ocean is one of the main reservoirs of sulfur, where it mainly occurs as dissolved SO_4^{2-} in seawater or as mineral in the form of pyrite (FeS_2) and gypsum ($CaSO_4$) in sediments (Sievert et al., 2007). Sulfur exists in different oxidation states, with sulfide (S^{2-}), elemental sulfur (S^0) and SO_4^{2-} as the most abundant and stable species in nature. With an amount of 29 mM, SO_4^{2-} is the most dominant anion in ocean water, next to chloride. The sedimentary sulfur cycle involves two main microbial processes: (i) bacterial dissimilatory reduction of SO_4^{2-} to hydrogen sulfide, which can subsequently precipitate with metal ions (mainly iron), and (ii) assimilatory reduction of SO_4^2 to form organic sulfur compounds incorporated in microbial biomass (Jørgensen & Kasten, 2006). Dissimilatory SO_4^{2-} reduction by SRB occurs in anoxic marine sediments or in freshwater environments, where SRB use several electron donors, such as hydrogen, various organic compounds (e.g. ethanol, formate, lactate, pyruvate, fatty acids, methanol, and methanethiol) as well as CH_4 (Muyzer & Stams, 2008).

AOM was considered impossible in the past, due to the non polar C-H bond of CH_4 (Thauer & Shima, 2008). From a thermodynamic point of view, AOM-SR yields minimal energy: only 17 kJ mol^{-1} (Eq. 2.1 in Figure 2.4). In comparison, more energy is released by the hydrolysis of one adenosine triphosphate (31.8 kJ mol^{-1}). Other electron acceptors in the anaerobic environment, such as nitrate, iron and manganese provide higher energy yields than SO_4^{2-}, as deducted by the $\Delta G^{0'}$ of the different redox reactions (Figure 2.4). However, their combined concentration at the marine sediment-water interface is far lower than the SO_4^{2-} concentration (D'Hondt et al., 2002). Therefore, AOM-SR usually dominates in marine sediments.

AOM-SR was suggested to be a cooperative metabolic process of the AOM coupled to dissimilatory SO_4^{2-} reduction, thereby gaining energy from a syntrophic consortium of ANME and SRB (Boetius et al., 2000; Hoehler et al., 1994) (Eq. 2.1 in Figure 2.4). Especially the *Desulfosarcina* / *Desulfococcus* (*DSS*) and *Desulfobulbaceae* (*DBB*) clades of SRB are common associates of ANME for SR. However, the three ANME phylotypes have been visualized without any attached SRB in different marine environments as well (Losekann et al., 2007; Maignien et al., 2013; Treude et al., 2005a; Wankel et al., 2012a), suggesting that AOM-SR can potentially be performed independently by the ANME themselves (Eq. 2.1 in Figure 2.4, performed solely by ANME). Theoretically, slightly more energy can be released (18 kJ mol^{-1}) if SO_4^{2-} is reduced to disulfide instead of sulfide (Milucka et al., 2012). Microorganisms that mediate AOM-SR can also use $S_2O_3^{2-}$ or S^0 sulfur as terminal electron acceptor for AOM (Eq. 2.7 and 2.8 in Figure 2.4). The reduction of one mole of $S_2O_3^{2-}$ to one mole sulfide requires fewer electrons (4 electrons) than the reduction of SO_4^{2-} to sulfide (8 electrons). Although the theoretical Gibbs free energy for AOM coupled to S^0 is positive (+24 kJ mol^{-1}, Eq. 2.7 in Figure 2.4), *in vitro* tests showed that this reaction may well proceed and the calculated free energy of reaction at *in situ* conditions is negative (-84.1 kJ mol^{-1}) (Milucka et al., 2012).

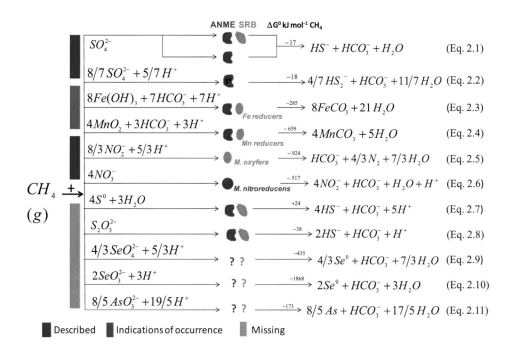

Figure 2.4 *Described and possible AOM processes with different terminal electron acceptors. The AOM with SO_4^{2-}, nitrate and nitrite as electron acceptor is well described along with the microbes involved, which is indicated by green block whereas the AOM with manganese and iron was shown but the microbes involved need to be characterized which is indicated by blue block. Other possible*

electrons are mentioned in the bottom part of the figure according to the thermodynamic calculation of the chemical reaction, which is indicated by orange block.

2.2.4 AOM coupled to nitrite and nitrate reduction

Methanotrophs that utilize nitrite (Ettwig et al., 2010) or nitrate (Haroon et al., 2013) have been identified in anaerobic fresh water sediments. Thermodynamically, the CH_4 oxidation coupled to nitrite and nitrate yields more energy than AOM-SR, with a $\Delta G^{0'}$ of -990 kJ mol^{-1} and -785 kJ mol^{-1}, respectively (Eq. 2.5 and 2.6 in Figure 2.4). Two specific groups of microbes are involved in the process of AOM coupled with nitrate and nitrite reduction: "*Candidatus Methanoperedens nitroreducens*" (archaea) and "*Candidatus Methylomirabilis oxyfera*" (bacteria), respectively.

AOM coupled to denitrification was first hypothesized to occur in a similar syntrophic manner as AOM-SR (Raghoebarsing et al., 2006). However, Ettwig et al. (2010) showed that CH_4 oxidation coupled to nitrite reduction occurs in the absence of archaea. The bacterium "*Candidatus Methylomirabilis oxyfera*" couples AOM to denitrification, with nitrite being reduced to nitric oxide which is then converted to nitrogen and oxygen. Thus generated intracellular byproduct oxygen is subsequently used to oxidize CH_4 to CO_2 (Ettwig et al., 2010). Moreover, recent studies reveal that a distinct ANME, affiliated to the ANME-2d subgroup and named "*Candidatus Methanoperedens nitroreducens*" (Figure 2.2C and 2.3), can carry out AOM using nitrate as the terminal electron acceptor through reversed methanogenesis (Haroon et al., 2013). In the presence of ammonium, the nitrite released by this ANME-2d is then reduced to nitrogen by the anaerobic ammonium-oxidizing bacterium (anammox) "*Candidatus Kuenenia spp.*", while in the absence of ammonium, nitrate is reduced to nitrogen by "*Candidatus Methylomirabilis oxyfera*". Therefore, different co-cultures are dominated in a biological system depending on the availability of the nitrogen species (nitrate, nitrite or ammonium) (Haroon et al., 2013).

2.2.5 Other electron acceptors for AOM

Besides SO_4^{2-} and nitrate, iron and manganese are other electron acceptors studied for AOM. In marine sediments, AOM was found to be coupled to the reduction of manganese or iron (Beal et al., 2009), but whether manganese and iron are directly used for the process or not, is yet to be elucidated. An *in vitro* study from Beal et al.

(2009) showed that oxide minerals of manganese, birnessite (simplified as MnO_2 in Eq. 2.4 of Figure 2.4) and iron, ferrihydrite (simplified as $Fe(OH)_3$ in Eq. 2.3 of Figure 2.4), can be used as electron acceptors for AOM. The rates of AOM coupled to MnO_2 or $Fe(OH)_3$ reduction are lower than AOM-SR, but the energy yields ($\Delta G^{0'}$ of -774 kJ mol^{-1} and -556 kJ mol^{-1} respectively, Eq. 2.3 and 2.4 in Figure 2.4) are higher. Thus, the potential energy gain of Mn- and Fe-dependent AOM is, respectively, 10 and 2 times higher than that of AOM-SR, inspiring researchers to further investigate these potential processes (Beal et al., 2009).

Several researchers have investigated the identity of the bacteria present in putative iron- and manganese-dependent AOM sites and hypothesized their involvement along with ANME (Beal et al., 2009; Wankel et al., 2012a). In parallel to AOM-SR, this process is also assumed to be mediated by two cooperative groups of microorganisms. Bacterial 16S rRNA phylotypes found in iron- and manganese-dependent AOM sites are putative metal reducers, belonging to the phyla *Verrucomicrobia phylotypes* (Wankel et al., 2012a), *Bacteriodetes*, *Proteobacteria* and *Acidobacteria* and are mostly present in heavy-metal polluted sites and hydrothermal vent systems (Beal et al., 2009). The ANME-1 clade was identified as the most abundant in metalliferous hydrothermal sediments and in Eel River Basin CH_4-seep sediment. However, the sole identification of specific bacteria and archaea in these marine sediments does not provide evidence for their metal reducing capacity.

Recent studies assumed the direct coupling of AOM to iron reduction. Wankel et al. (2012a) investigated AOM in hydrothermal sediments from the Middle Valley vent field, where AOM occurred in the absence of SR and SRB. Fe-dependent AOM was hypothesized as the process in these sediments, due to the abundance of iron bearing minerals, specifically green rust and a mixed ferrous-ferric hydroxide. A higher AOM rate than with SR was observed in *in vitro* incubations with mangenese and iron based electron acceptors like birnessite and ferrihydrite (Segarra et al., 2013).

There is also a hypothesis on possible indirect coupling of AOM with metal reduction (Beal et al., 2009). Namely, sulfide, present in the sediment, is oxidized to S^0 and disulfide in the presence of metal oxides. The produced sulfur compounds can be disproportionated by bacteria producing transient SO_4^{2-}, which can be used to oxidize CH_4. These sulfur transformations are referred to as cryptic sulfur cycling (Aller & Rude, 1988; Canfield et al., 1993) and its extent can be increased if the sediment is rich in microorganisms able to metabolize S^0 and disulfide (Straub & Schink, 2004; Wan et al., 2014). A recent study with the Bothnian Sea sediment speculated two separate anaerobic regions where AOM occurs: AOM-SR (in the upper anaerobic layer) and iron dependent AOM (in the lower anaerobic layer). It was hypothesized that the majority of AOM was coupled directly to iron reduction in the iron reducing region and only about 0.1% of AOM-SR was due to cryptic sulfur cycling (Egger et al., 2015). However, in marine and brackish sediments probably only a few percent of the CH_4 is oxidized by a iron-dependent process.

Theoretically, based on thermodynamics, anaerobic CH_4 oxidizing microorganisms could utilize other electron acceptors including the oxyanions of arsenic and selenium. It should be noted that the chemistry of selenium oxyanions is similar to that of sulfur oxyanions, since both belong to the same group in the periodic table, the so called chalcogens. Oxidized selenium species i.e., selenate or selenite, might thus also be used as electron acceptor for AOM (Eq. 2.9 and 2.10 in Figure 2.4).

2.3 Physiology of ANME

2.3.1 Carbon and nitrogen metabolism

The difficulty in obtaining enrichment cultures of ANME hampers getting insights into the physiological traits of these microorganisms. Nonetheless, *in situ* and *in vitro* activity tests using ^{13}C- or ^{14}C-labelled CH_4 unequivocally revealed that ANME oxidize CH_4 (Nauhaus et al., 2007). But the physiology of these microorganisms seems to be more intriguing. Recently, it was found that the carbon in ANME biomass is not totally derived from CH_4, i.e., ANME are not obligate heterotrophs. ANME-2 seems to assimilate carbon from CH_4 as well as from CO_2 at similar amounts (Wegener et al., 2008), whereas carbon within the biomass of ANME-1 is derived from CO_2 fixation (Kellermann et al., 2012; Treude et al., 2007). Thus, these ANME are regarded as CH_4-oxidizing chemoorganoautotrophs (Kellermann et al., 2012). Furthermore, genetic studies showed that ANME-1 contains genes encoding the CO_2 fixation pathway characteristic for methanogens (Meyerdierks et al., 2010).

There is evidence that some ANME-1 and/or ANME-2 from the Black Sea and from the Gulf of Mexico CH_4 seeps can produce CH_4 (Orcutt et al., 2005; Treude et al., 2007) from CO_2 or from methanol (Bertram et al., 2013). This methanogenic capacity exhibited by these ANME seems in turn to mirror the CH_4 oxidation capacity displayed by pure cultures of methanogens (Harder, 1997; Zehnder & Brock, 1979) and by methanogens present in anaerobic sludge (Meulepas et al., 2010b), which can oxidize about 1 to 10% of the CH_4 they produce. However, the reported CH_4 oxidation capacity of cultured methanogens is so low that they are not considered to contribute to CH_4 oxidation in marine settings. On the contrary, the detection of important numbers of active ANME-1 cells in both the CH_4 oxidation and the CH_4 production zones of estuary sediments has led to the proposition that this ANME type is not an obligate CH_4 oxidizer, but rather a flexible type which can switch and function as methanogen as well (Lloyd et al., 2011).

Another intriguing physiological trait is the nitrogen fixing capacity (i.e., diazotrophy) by ANME-2d. Using $^{15}N_2$ as nitrogen source, it was found that ANME-2d cells assimilated ^{15}N in batch incubations of marine mud volcano or CH_4 seep sediments (Dekas et al., 2014; Dekas et al., 2009). While fixing nitrogen, ANME maintained their CH_4 oxidation rate, but their growth rate was severely reduced. The energetic cost to fix nitrogen is one of the highest amongst all anabolic processes and requires about 16 ATP molecules, which translates into 800 kJ mol^{-1} of nitrogen reduced. Therefore, considering the meager energy gain of AOM (about 18 kJ mol^{-1} of CH_4 oxidized), it is consistent that the growth rate of ANME can be 20 times lower using nitrogen than using ammonium (NH_4^+) as nitrogen source (Dekas et al., 2009). Yet, it is not resolved under which *in situ* conditions these microorganism would be diazotrophic. Also, whether other ANME types are diazotrophs has not yet been shown. Although the metagenome of ANME-1 reveals the presence of various candidate proteins having similarity to proteins known to be involved in nitrogen fixation (Meyerdierks et al., 2010), this trait has not yet been tested experimentally for fixing nitrogen.

2.3.2 Syntrophy and potential electron transfer modes between ANME and SRB

Several theories have been proposed to understand the mechanism between ANME archaea and their association with SRB, with the most common hypothesis of syntrophy between ANME and SRB (Figure 2.5A). Obligate syntrophs share the substrate degradation process resulting in one partner converting the substrate into an intermediate, which is consumed by the syntrophic partner (Stams & Plugge, 2009). Unlike other known forms of syntrophy, the intermediate shared by ANME and SRB has not yet been identified. The syntrophy between ANME and SRB is hypothesized on the basis of the tight co-occurrence of ANME and SRB in AOM active sites, as revealed by FISH images (Figures 2.2B and 2.2D) (Blumenberg et al., 2004; Boetius et al., 2000; Knittel et al., 2005). Isotopic signatures in archaeal and bacterial lipid biomarker based analysis strengthened this hypothesis, assuming transfer of an intermediate substrate between the two microorganisms (Boetius et al., 2000; Hinrichs & Boetius, 2002; Hinrichs et al., 2000). Also phylogenetic analysis showed the co-occurrence of SRB and ANME in samples from AOM sites, suggesting that their co-existence may play a role in the AOM (Alain et al., 2006; Losekann et al., 2007; Stadnitskaia et al., 2005).

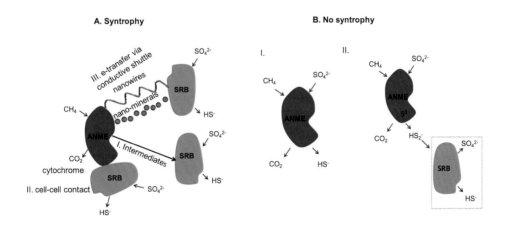

Figure 2.5 *Syntrophic and non-syntrophic ANME using SO_4^{2-} as an electron acceptor. A. In a syntrophic association, ANME can transfer electrons to a SRB via different mechanisms: i) electron transfer via possible intermediate compounds such as, formate, acetate or hydrogen, ii) electron transfer via cell to cell contact between ANME and SRB through cytochromes, and iii) electron transfer via nanowires or nanominerals acting like a conduit between the ANME and SRB. B. In a non-syntrophic association: i) ANME can possibly perform the complete AOM process alone with CH_4 oxidation and SO_4^{2-} reduction without a SO_4^{2-} reducing partner or ii) ANME can perform AOM alone by producing CO_2 and disulfide (HS_2^-) with S^o as an intermediate. The HS_2^- is disproportionated by SRB (Milucka et al., 2012).*

Hydrogen and other methanogenic substrates, such as acetate, formate, methanol and methanethiol were hypothesized as the intermediates between ANME and SRB (Figure 2.5A) (Hoehler et al., 1994; Sørensen et al., 2001; Valentine et al., 2000). Formate is the only possible intermediate which would result in free energy gain, so thermodynamic models support formate as an electron shuttle (e-shuttle) of AOM (Sørensen et al., 2001). However, acetate was assumed to be the favorable e-shuttle in high CH_4 pressure environments (Valentine, 2002). Genomic studies suggested that the putative intermediates for AOM could be acetate, formate or hydrogen (Hallam et al., 2004; Meyerdierks et al., 2010). Therefore, acetate could be formed by ANME-2a and be a possible intermediate. Considering AOM as a reversed methanogenesis, the first step is the conversion of CH_4 to methyl-CoM and the pathway involves the production of either acetate or hydrogen as an intermediate (Hallam et al., 2004; Wang et al., 2014). Nevertheless, the addition of hydrogen in an AOM experiment does not illustrate any change in AOM rate, in contrast to the typical methanogenesis process (Moran et al., 2008). Similarly, CH_4 based SR rates were the same even if these potential intermediates (acetate/formate) were supplied, whereas the reaction should be shifted to lower AOM rates upon the addition of intermediates (Meulepas et al., 2010a; Moran et al., 2008). Moreover, the addition of these potential intermediates induces the growth of different SRB than the *DSS* and *DBB* groups, which are the assumed syntrophic partner of ANME (Nauhaus et al., 2005). Therefore, the hypothesis of these compounds being possible AOM e-shuttles is unconfirmed.

The formate dehydrogenase gene is highly expressed in the ANME-1 genome, thus formate can be formed by ANME-1 and function as intermediate (Meyerdierks et al., 2010). Likewise, the ADP-forming acetyl-CoA synthetase which converts acetyl-CoA to acetate was retrieved in the ANME-2a genome (Wang et al., 2014). Instead, methyl sulfide was proposed to be an intermediate for both methanogenesis and methanotrophy (Moran et al., 2008). Methyl sulfide is then assumed to be produced by the ANME and can be utilized by the SRB partner (Moran et al., 2008).

In case of no evidence for intermediates, a possible e-shuttling mechanism between the ANME and SRB could be the transfer of electrons via nanowires, nanominerals or via direct cell to cell contact with electron transfer (e-transfer) via cytochromes (Figure 2.5A), but also these hypothesis still need to be proven (Meyerdierks et al.,

26

2010; Shima & Thauer, 2005; Stams & Plugge, 2009; Wang et al., 2014). Wegner et al. (2008) showed the uptake of inorganic carbon by SRB into its fatty acid, suggesting the SRB as autotrophic. Nevertheless, there was no proof for direct CH_4 consumption by the SRB. In such case, the more favorable e-shuttle could be e-transfer by a difference in redox in the cell membrane through cell to cell contact or via cytochromes (Figure 2.5A) (Wegener et al., 2008). Multiheme c-type cytochromes were identified in the ANME-1 archaea genome, so the direct membrane bound e-transfer between ANME-1 and SRB could be possible (Meyerdierks et al., 2010). The c-type cytochrome specific gene was also well expressed in the ANME-2a according to a metatranscriptome study (Wang et al., 2014). The importance of multiheme c-type cytochromes has been extensively discussed in Geobacter species, where the cytochrome can act as an electron storage in the cell membrane and subsequent extracellular e-transfer occurs (Lovley, 2008).

Another mode of e-transfer can occur via highly conductive pili or nanowires (Figure 2.5A), which have been well described in the Fe(III) reducer Geobacter sulfurreducens (Reguera et al., 2005). Recently, a longer nanowire, which facilitates the e-transfer in the range of centimeter scales, has been described in marine environments (Nielsen et al., 2010; Vasquez-Cardenas et al., 2015). A novel cable bacterium putatively belonging to the Desulfobulbaceae and growing as a long filament of around 1 cm length performing e-transport was retrieved from the marine lake Grevelingen (North Sea, The Netherlands) (Vasquez-Cardenas et al., 2015). These cable bacteria are heterotrophs, yet the carbon metabolism has to be defined. Together with other chemolithoautotrophic bacteria, cable bacteria perform electrogenic sulfur oxidation via their long filament (Vasquez-Cardenas et al., 2015). Interestingly, one clade of the cocci shaped DBB is commonly associated with ANME (specifically ANME-3), however long filaments have not yet been visualized in the FISH based studies of ANME sites (Niemann et al., 2006b).

Nanominerals can also be a possible mode of interspecies e-transfer in anaerobic microbial communities (Figure 2.5A). This mode of e-transfer was described between Geobacter sulfureducens and Thiobacilus denitrificans for acetate oxidation with nitrate reduction (Kato et al., 2012b). Iron oxide nanominerals (10-20 nm diameter) resulting from the microbial oxidation and reduction of iron act as electron transporter, which receives the electrons from one cell and discharges them to other cells. These conductive iron oxide nanominerals are also assumed to facilitate methanogenesis (Kato et al., 2012a). When the iron oxide nanomineral was supplied to the methanogens from rice paddy soil, the methanogenesis rates were increased (Kato et al., 2012a). Recently, addition of iron oxide was also shown to facilitate the AOM-SR in CH_4 seeps (Sivan et al., 2014), however the role of these nanominerals as possible interspecies e-shuttle among ANME and SRB has to be further investigated.

2.3.3 Non-syntrophic growth of ANME

The syntrophic requirement of ANME conducting AOM-SR is a topic of debate (Boetius et al., 2000; Hinrichs et al., 1999; Maignien et al., 2013; Wankel et al., 2012a). Different carbon compounds could be utilized as intermediates between the archaea and bacteria or direct/indirect cell contact could be the preferred e-transfer mode. However, recent research shows that a certain ANME phylotype may conduct AOM-SR without the need of SRB with S^0 and/or disulfides as the end product (Milucka et al., 2012). AOM-SR generates little energy, sharing this energy between two microorganisms in a syntrophic association may thus not be energetically as advantageous as conducting the process in a non-syntrophic mode. Visualization of ANME and its bacterial partners by FISH showed that for all three clades of ANME, the association with SRB is not obligatory. The recent view is that, in some cases, the AOM process could occur by only the ANME without any SRB, especially for ANME-1 (Wankel et al., 2012a) and ANME-2 (Milucka et al., 2012) (Figure 2.5B). The possibility of non-syntrophic growth of ANME is further supported by the presence of nickel containing methyl-coenzyme M reductase (MCR) in ANME-1 and ANME-2, like other methanogens (Hallam et al., 2004; Wang et al., 2014). Scheller et al. (2010) discussed the MCR is able to break the stable C-H bond of CH_4 without any involvement of highly reactive oxidative intermediates.

In some habitats and microbial mats only the single shell type ANME was observed, such as in the pink microbial mat from the Black Sea (90% ANME-1) (Michaelis et al., 2002) and ANME-1 in metalliferous hydrothermal vents (Wankel et al., 2012a). Mostly, ANME-1 archaea exist as single cells or as monospecific chains without any attached partner (Maignien et al., 2013; Orphan et al., 2002). AOM-SR occurring in high temperature sediments is also dominated by ANME-1, showing the decoupling of AOM with SR (Wankel et al., 2012a). ANME-2 were visualized without bacterial partner even though SRB were abundant on the site (Treude et al., 2005b), while ANME-3 were detected as monospecific aggregates or as aggregates with unidentified bacteria (Losekann et al., 2007). In these cases, the non association with SRB has been justified by their growth environment or the use of electron acceptors other than SO_4^{2-}. For instance, ANME-1 are found more abundant in deeper sediments where the SO_4^{2-}. concentration is lower and the concentration of iron oxides is higher, suggesting that other electron acceptors might be involved instead of SO_4^{2-} (Knittel et al., 2005; Omoregie et al., 2009; Orcutt et al., 2005).

A recent study showed a completely different view about the non-syntrophic behavior of ANME (Milucka et al., 2012). A new AOM mechanism was revealed, in which ANME were responsible for both AOM and SR (Figure 2.5B). CH_4 was oxidized to bicarbonate and then the SO_4^{2-} was reduced to S^0, as an intracellular intermediate in ANME-2 cells. The resulting S^0 was then released outside the cell as disulfide, which is converted to HS^- by the SRB. Figure 2.5B shows some ANME can sustain the overall AOM reaction without bacterial partner, even though the *DSS* type *Deltaproteobacteria* render the AOM-SR more thermodynamically favorable by scavenging the disulfide by disproportionation or dissimilatory reduction. The disulfide produced by ANME is disproportionated into SO_4^{2-} and sulfide. The thus produced SO_4^{2-} can be used again by the ANME, while sulfide can undergo several conversions, for instance precipitate as FeS_2 or partially (to S^0) or completely (to SO_4^{2-}) oxidize aerobically or anaerobically (in the presence of light by e.g. purple sulfur bacteria) (Dahl & Prange, 2006). As described earlier, in the presence of iron oxides, sulfide can react abiotically forming more substrates (disulfide and S^0) for the *Deltaproteobacteria*. The reaction of sulfide with iron oxides can thus strongly enhance the sulfur cycle, similarly to the study conducted with *Sulfurospirillum deleyianum* (Straub & Schink, 2004).

2.4 Drivers for distribution of ANME in natural habitats

2.4.1 Major habitats of ANME

ANME are widely distributed in marine habitats including cold seep systems (gas leakage from methane hydrates), hydrothermal vents (fissures releasing hot liquid and gas in the seafloor) and organic rich sediments with diffusive CH_4 formed by methanogenesis (Figure 2.6 and Figure 2.7). The cold seep systems include mud volcanoes, hydrate mounds, carbonate deposits and gaseous carbonate chimneys (Boetius & Wenzhöfer, 2013), which are all frequently studied ANME habitats. The major controlling factors for the ANME distribution are the availability of CH_4 and SO_4^{2-} or other terminal electron acceptors which can possibly support the AOM, whilst other environmental parameters such as temperature, salinity, and alkalinity also play a decisive role in ANME occurrence. Among the three clades, ANME-2 and ANME-3 apparently inhabit cold seeps, whereas ANME-1 is cosmopolitan, residing in a wide temperature and salinity range (Table 2.1 and Figure 2.7). Recently, AOM has been reported in non-saline and terrestrial environments as well, for instance in the Apennine terrestrial mud volcanoes (Wrede et al., 2012) and in the Boreal peat soils of Alaska (Blazewicz et al., 2012) (Table 2.2).

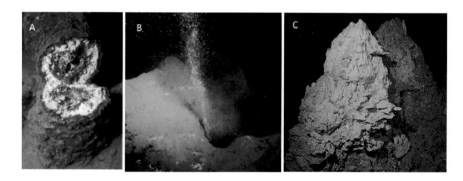

Figure 2.6 *In situ pictures of some of the well studied ANME habitats. A) Giant microbial mat in carbonate chimney the Black Sea (Blumenberg et al., 2004), B) CH4 bubble seeping from Haakon Mosby mud volcano and C) Carbonate chimney from the Lost City hydrothermal vent (Brazelton et al., 2006)*

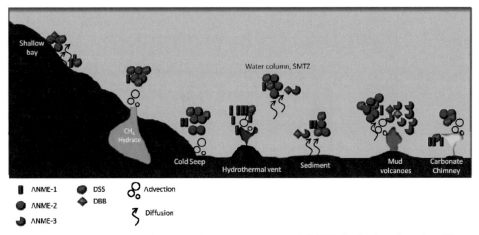

Figure 2.7 *Major habitats of ANME in marine environments and ANME distribution along the different major habitats. ANME-1 is mainly inhabited in diverse environments including hydrothermal vents, cold seeps and carbonates chimneys whereas ANME-2 was retrieved from diverse cold seeps, CH_4 hydrates and mud volcanoes and ANME-3 was mainly retrieved from a specific mud volcano. ANME types: ($\blacksquare$) ANME-1, ($\bullet$) ANME-2 and ($\bullet$) ANME-3. SRB types: DSS ($\bullet$) and DBB ($\blacklozenge$).CH_4 transport regime: advection ($\mathbf{O}$) and diffusion ($\langle$).*

The Black Sea, a distinct ANME habitat, consists of thick microbial mats of ANME-1 and ANME-2 (2-10 cm thick) adhered with carbonate deposits (chimney-like structure) in various water depths of 35-2000 m (Blumenberg et al., 2004; Michaelis et al., 2002; Novikova et al., 2015; Reitner et al., 2005; Thiel et al., 2001; Treude et al., 2005a). CH_4 is distributed by vein like capillaries throughout these carbonate chimneys and finally emanated to the water column (Krüger et al., 2008; Michaelis et al., 2002; Treude et al., 2005a). These microbial habitats are of different size and nature, such as small preliminary microbial nodules (Treude et al., 2005a), floating microbial mats (Krüger et al., 2008) and large chimneys (Michaelis et al., 2002). The immense carbonate chimney from the Black Sea (Figure 2.6A), with up to 4 m height and 1 m width, was marked to harbor an ANME-1 dominant pink-colored microbial mat with the highest known AOM rates in natural systems (Blumenberg et al., 2004; Michaelis et al., 2002). Deep sea carbonate deposits from cold seeps and hydrates are active and massive sites for AOM and ANME habitats (Marlow et al., 2014b).

Table 2.1 Rates of AOM and SR in different natural marine habitats along with dominant ANME types. Here the depth refers to water depth. The different methods of AOM and SR measurements are indicated by the superscript letters next to the references as follows: [a]= in situ measurement, [b]= ex situ radiotracer measurement, [c]= model calculation, [d]= pore water chemistry measurement

Location	Depth (m)	CH_4 (mM)	SO_4^{2-} (mM)	ANME types	AOM rates	SR rates	References
Cold seeps (temperature ranging from 1.5 to 20°C)							
Black Sea (giant carbonate chimney)	230	2.8	17	ANME-1	7800 to 21000 nmol g_{dw}^{-1} d^{-1}	4300 to 19000 nmol g_{dw}^{-1} d^{-1}	(Michaelis et al., 2002; Treude et al., 2007)[a]
Black Sea (other microbial mats)	180	3.7	9 to 15	ANME-1, ANME-2	2000 to 15000 nmol g_{dw}^{-1} d^{-1}	4000-20000 nmol g_{dw}^{-1} d^{-1}	(Krüger et al., 2008)[a]
Haakon Mosby mud volcano, Barents Sea	1250	0.0003 to 0.0057	-	ANME-2, ANME-3	1233 to 2000 nmol cm^{-2} d^{-1}	2250 nmol cm^{-2} d^{-1}	(Niemann et al., 2006b)[b]
Gulf of Mexico, hydrate	550 to 650	2 to 6	20	ANME-1, ANME-2	280±460 nmol cm^{-2} d^{-1}	5400±9400 nmol cm^{-2} d^{-1}	(Joye et al., 2004; Orcutt et al., 2005)
Eel River Basin carbonate mounds and hydrates	500 to 850	3	20	ANME-1, ANME-2	200 nmol cm^{-3} d^{-1}	-	(Marlow et al., 2014b; Orphan et al., 2004)[b]
Gulf of Cadiz, mud volcanoes	810 to 3090	0.001 to 1.3	10 to 40	ANME-2, ANME-1	10 to 104 nmol cm^{-2} d^{-1}	158 to 189 nmol cm^{-2} d^{-1}	(Niemann et al., 2006a)[b]

Sample			ANME			Reference
Black Sea water	100 to 1500	0.011	-	ANME-1, ANME-2	0.03 to 3.1 nmol d^{-1}	-	(Durisch-Kaiser et al., 2005; Schubert et al., 2006)[b]
Tommeliten seepage area, North Sea sediment	75	1.4 to 2.5	30 to 20	ANME-1, ANME-2	1.4 to 3 nmol cm^{-3} d^{-1}	3 to 4.6 nmol cm^{-3} d^{-1}	(Niemann et al., 2005)[b]
Methane rich sediments (temperature from 4 to 20°C)							
Bothnian Sea sediment	200	2	5.5	-	40 to 90 nmol cm^{-2} d^{-1}	-	(Slomp et al., 2013)[c]
Baltic Sea/ Eckernförde Bay sediment	25	0.001 to 0.8	16 to 21	ANME-2	1 to 14 nmol cm^{-3} d^{-1}	20 to 465 nmol cm^{-3} d^{-1}	(Treude et al., 2005b)[b]
Skagerrak sediment	308	1.3	25	ANME-2, ANME-3	3 nmol cm^{-3} d^{-1}	-	(Parkes et al., 2007)[d]
West African margin sediment	400 - 2200	1 to 19	26	-	0.0027 nmol cm^{-3} d^{-1}	-	(Sivan et al., 2007)[c]
Hydrothermal vents (temperature from 10 to 100°C)							
Guaymas Basin hydrothermal vent	-	-	-	ANME-1	1200 nmol g$_{dw}$$^{-1}$ d^{-1} | 250 nmol g$_{dw}$$^{-1}$ d^{-1}	(Holler et al., 2011b)[a]
Juan de Fuca Ridge hydrothermal vent	2400	3	-	ANME-1	11.1 to 51.2 nmol cm^{-3} d^{-1}	-	(Wankel et al., 2012a)[c]

Table 2.2 *Rates of AOM and SR in different natural terrestrial habitats. The different methods of AOM and SR measurements are indicated by the superscript letter next to the reference as follows:*

[a]= in vitro measurement, [b]= ex situ radiotracer measurement

Location	Soil depth (cm)	CH_4 (mM)	SO_4^{2-} (mM)	AOM rates	SR rates	References
Wetland and peat soil	0-40	0.5-1	0.1-1	265 ± 9 nmol cm^{-3} d^{-1}	300 nmol cm^{-3} d^{-1}	(Segarra et al.,2015)[b]
Tropical forest soil, Alaska	10-15	-	< 1	3-21 nmol g$_{dw}$$^{-1}$ d^{-1}	-	(Blazewicz et al., 2012)[a]
Paclele Mici mud volcano, Carpathian mountains			1.5-2	2-4 nmol g^{-1} d^{-1}	-	(Alain et al., 2006)[a]
Peat land soil from diverse places	30-50			0.03-3 nmol g^{-1} d^{-1}	-	(Gauthier et al., 2015)[a]

Likewise, CH_4 based authigenic carbonate nodules and CH_4 hydrates which host ANME-1 and ANME-2 (Marlow et al., 2014b; Mason et al., 2015; Orphan et al., 2001a; Orphan et al., 2001b; Orphan et al., 2002) prevail in the Eel River Basin (off shore California), a cold seep with an average temperature of 6°C and known for its gas hydrates (Brooks et al., 1991; Hinrichs et al., 1999). Both ANME-1 and ANME-2 are commonly associated with *DSS* in the sediments of Eel river, however ANME-1 appeared to exist as single filaments or monospecific aggregates in some sites as well (Hinrichs et al., 1999; Orphan et al., 2001b; Orphan et al., 2002).

Other cold seep sediments were also extensively studied as ANME habitats. The Gulf of Mexico, a cold seep with bottom water temperature of 6°C to 8°C, is known for its gas seepage and associated hydrates. These CH_4 hydrates located at around 500 m seawater depth in the Gulf of Mexico are inhabited by diverse microbial communities: *Beggiatoa* mats with active AOM are common bottom microbial biota in the sulfidic sediments (Joye et al., 2004; Lloyd et al., 2006; Orcutt et al., 2005; Orcutt et al., 2008). ANME-1 dominates the sediment of the Gulf of Mexico, particularly in the hypersaline part as a monospecific clade, whereas ANME-2 (a and b) are present together with *DSS* groups in the less saline hydrates (Lloyd et al., 2006; Orcutt et al., 2005). Similarly, different mud volcanoes of the Gulf of Cadiz cold seep harbor ANME-2 with the majority being ANME-2a (Niemann et al., 2006a), whereas the hypersaline Mercator mud volcano of the Gulf of Cadiz hosts ANME-1 (Maignien et al., 2013). Retrieval of ANME-1 in the hypersaline environment suggests the ANME-1 adaptability to wider salinity ranges compared to other ANME phylotypes. Mud volcanoes from the Eastern Mediterranean (Kazan and Anaximander mountains) are inhabited by all three ANME phylotypes, whereas Kazan mud volcano hosts the distinct ANME-2c clade (Heijs et al., 2007; Kormas et al., 2008; Pachiadaki et al., 2010; Pachiadaki et al., 2011). Likewise, Haakon Mosby mud volcano in the Barents Sea is the firstly described habitat for ANME-3 with almost 80% of the microbial cells being ANME-3 and *DBB* (Figure 2.6B and Figure 2.2D) (Losekann et al., 2007; Niemann et al., 2006b).

Some of the hydrothermal vents are well studied ANME habitats for distinct ANME clades and thermophilic AOM. The Guaymas Basin in the California Bay, an active hydrothermal vent with a wide temperature range, is known for the occurrence of different ANME-1 phylotypes, along with unique thermophilic ANME-1 (Biddle et al.,

2012; Larowe et al., 2008; Vigneron et al., 2013). ANME-1 is predominant throughout the Guaymas Basin, yet the colder CH_4 seeps of the Sonara Margin host all three ANME phylotypes (ANME-1, ANME-2 and ANME-3) with peculiar ANME-2 (ANME-2c Sonara) (Vigneron et al., 2013). Likewise, mesophilic to thermophilic AOM carried out by the ANME-1 clade was detected in the Middle Valley vent field on the Juan de Fuca Ridge (Lever et al., 2013; Wankel et al., 2012a). Another vent site, the Lost City hydrothermal vent with massive fluid circulation and ejecting hydrothermal fluid of > 80°C predominantly hosts ANME-1 within the calcium carbonate chimneys (Figure 2.6C), which are very likely deposited due to bicarbonate formation from AOM (Bradley et al., 2009; Brazelton et al., 2006).

2.4.2 ANME types distribution by temperature

The ANME clades exhibit a distinct pattern of distribution according to the temperature. ANME-2 and ANME-3 seem more abundant in cold seep environments, including hydrates and mud volcanoes, with an average temperature of 2 to 15°C. In contrast, ANME- 1 is more adapted to a wide temperature range from thermophilic conditions (50-70°C) to cold seep microbial mats and sediments (4-10°C) (Holler et al., 2011b; Orphan et al., 2004). Temperature appears to control the abundance of the ANME clades. However, some of the ANME types (ANME-1ab) exhibit adaptability to a wide range of temperatures. Other geochemical parameters as salinity, CH_4 concentration and pressure can act together with temperature as selection parameters for the distribution of ANME in natural environments. ANME-1 was extensively retrieved across the temperature gradient between 2°C to 100°C in the Guaymas Basin from the surface to deep sediments (Teske et al., 2002). A phylogenetically distinct and deeply branched group of the ANME-1 (ANME-1GBa) was found in the high temperature Guaymas Basin hydrothermal vent (Biddle et al., 2012) and other geologically diverse marine hydrothermal vents such as the diffuse hydrothermal vents in Juan de Fuca Ridge in the Pacific Ocean (10-25°C) (Merkel et al., 2013). The thermophilic trait of ANME-1GBa is supported by its GC (guanine and cytosine) content in its 16S rRNA genes, as it holds a higher GC percentage (> 60 mol%) compared to other ANME types. The GC content is positively correlated with the optimum temperature of microbial growth, the elevated GC content of ANME-1GBa suggests ANME-1 GBa being a thermophilic microbial cluster, with on optimum growth temperature of 70°C or above (Merkel et al., 2013). Moreover, when the Guaymas ANME community was enriched *in vitro*, the highest AOM rate was obtained in the range of 45-60°C, indicating that the major community consists of thermophilic ANME-1 (Holler et al., 2011b).

Other ANME-1 phylotypes (ANME-1a and ANME-1b) were observed in wide temperature ranges (3°C to > 60°C) (Biddle et al., 2012). ANME-1a and ANME-1b were retrieved from different hydrothermal vent areas and cold seeps, for example the Guaymas Basin hydrothermal vent at > 60°C (Biddle et al., 2012), Lost City hydrothermal vent (Brazelton et al., 2006), the Sonora Margin cold seep of the Guaymas Basin (3°C) (Vigneron et al., 2013), mud volcanoes in the Eastern Mediterranean cold seep (14 - 20°C) (Lazar et al., 2012), the Gulf of Mexico (6°C) (Lanoil et al., 2001; Lloyd et al., 2006), Black Sea microbial mat and water column (8°C) (Knittel et al., 2005; Schubert et al., 2006) and Eel River Basin (6°C) (Hinrichs et al., 1999; Orphan et al., 2001b). The occurrence of ANME-1a and ANME-1b in cold seep environments suggests ANME-1a and ANME-1b to be putative mesophiles to psychrophiles. The GC percentage of 16S rRNA genes of ANME-1a and ANME-1b is around 55 mol%, which is common for mesophiles (Merkel et al., 2013).

In contrast, ANME-2 and ANME-3 have a narrow temperature range. ANME-2 clades (2a, 2b and 2c) appear predominant in marine cold seeps and in some SMTZ where the temperature is about 4-20°C. The major cold seep environments inhabited by ANME-2 are described in the previous section (see section: *Major habitats of ANME*). The adaptability of ANME-2 in the cold temperature range is also substantiated by bioreactor enrichments with Eckernförde Bay sediment, where the maximum AOM rate was obtained when the bioreactor was operated at 15°C rather than at 30°C, for ANME-2a (Meulepas et al., 2009a). Similarly, Eckernförde Bay *in vitro* AOM rate measurements showed a steady increment in AOM rates from 4°C to 20°C and subsequently decreased afterwards (Treude et al., 2005b). Conversely, the recently described clade ANME-2d affiliated *"Candidatus Methanoperedens nitroreducens"* (Figure 2.3C), which was enriched from a mixture of freshwater sediment and wastewater sludge (Haroon et al., 2013), grows optimally at mesophilic temperatures (22-35°C) (Hu et al., 2009).

ANME-3 is also known to be thriving in cold temperature environments including cold seeps and mud volcanoes. The ANME-3 clade was firstly retrieved from the Haakon Mosby mud volcano with a temperature of about -1.5°C (Niemann et al., 2006b). Later, ANME-3 was found in other cold seep areas as well, such as the Eastern Mediterranean seepages at about 14°C (Heijs et al., 2007; Pachiadaki et al., 2010) and the Skagerrak seep (Denmark, North Sea) at around 6-10°C (Parkes et al., 2007).

2.4.3 Methane supply mode as driver for distinct distribution of ANME

In some seafloor ecosystems, CH_4 is transported by diffusion due to concentration gradients. Diffusion dominated ecosystems are typically quiescent sediments. In contrast, in seafloor ecosystems with CH_4 seeps, CH_4 is transported by advection of CH_4-rich fluids. Due to the complex dynamics of CH_4 transport in advection dominated environments, estimations of *in situ* CH_4 oxidation rates by geochemical mass balances is rather difficult (Alperin & Hoehler, 2010). Based on *ex situ* tests, the AOM rates are higher in ecosystems where high CH_4 fluxes are sustained by advective transport than in diffusion dominated ecosystems (Boetius & Wenzhöfer, 2013). The velocity of the CH_4-rich fluid may result in an order of magnitude difference in AOM rates. Higher AOM rates were observed at sites with higher flow velocity (Krause et al., 2014), probably high flows of CH_4-rich fluid support dense ANME populations.

The extent of CH_4 flux and the mode of CH_4 transport (advection vs. diffusion) are certainly important drivers for ANME population dynamics. Mathematical simulations illustrate that the transport regime can control the activity and abundance of AOM communities (Dale et al., 2008). Multivariate and cluster analysis shows that the mode of CH_4 transport can possibly control AOM communities (Figure 2.8). CH_4-rich upward fluid flow at active seep systems restricts AOM to a narrow subsurface reaction zone and sustains high CH_4 oxidation rates. In contrast, pore water CH_4 transport dominated by molecular diffusion leads to deeper and broader AOM zones, which are characterized by much lower rates and biomass concentrations (Dale et al., 2008). In this context, Roalkvam et al. (2012) found that the CH_4 flux largely influenced the specific density of ANME populations. However, whether distinct

38

ANME types preferentially inhabit environments dominated by advective or diffusive CH_4 transport is not yet clear.

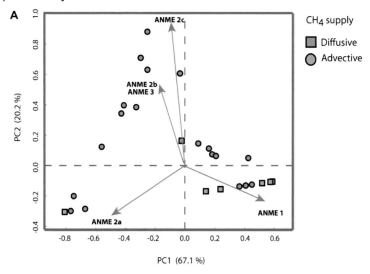

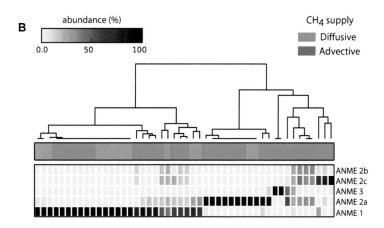

Figure 2.8 *The mode of CH_4 transport is apparently one of the drivers for the distribution of ANME types in the environment. A) Multivariate and B) cluster analyses show that ANME-2 is dominant mostly in CH_4-advective sites.*

At sites with high seepage activity like the Hydrate Ridge in Oregon, ANME-2 was dominant, whereas ANME-1 apparently was more abundant in the low seepage locations (Marlow et al., 2014a). A rough estimate of the abundance of the ANME type populations, reported in various marine environments, shows that ANME-2 dominate sites where CH_4 is transported by advection, while ANME-1 may dominate sites where CH_4 is transported by diffusion or advection (Figure 2.8). It is advisable that future studies regarding ANME type's distribution explicitly indicate the dominant mode of *in situ* CH$_4$ transport.

2.5 Ex situ enrichment of ANME

2.5.1 Need for enrichment of ANME

Molecular based methods allow the recognition of the phylogenetic diversity of ANME microorganisms in a wide range of marine sediments and natural environments. Determination of their detailed physiological and kinetic capabilities requires, until now, the cultivation and isolation of the microorganisms. The culturability of microorganisms inhabiting seawater (0.001 - 0.1%), seafloor (0.00001 - 0.6%) and deep subsea (0.1%) sediments is among the lowest compared to other ecosystems (Amann et al., 1995; D'Hondt et al., 2004; Parkes et al., 2000). This also holds for ANME from all thus far known environments, which so far have not yet been cultivated in pure culture for various reasons, not all known.

Specifically for the enrichment of ANME, the following aspects limit their cultivation: (i) from all known microbial processes, the AOM reaction with SO$_4^{2-}$ is among those which yield the lowest energy, (ii) the growth rate of these microorganisms is thus very low with a yield of 0.6 g$_{cell\ dw}$ mol^{-1} of CH$_4$ oxidized (Nauhaus et al., 2007), (iii) the dissolved concentrations of their substrate CH$_4$ (1.4 mM) at atmospheric pressures is limited to values far much lower than the estimated apparent half affinity constant for CH$_4$ (37 mM) during the AOM process and (iv) sulfide, which is a product of the reaction, can be inhibitory. All these aspects set a great challenge for the cultivation and isolation of ANME.

It is recognized that culturability can be enhanced when the conditions used for cultivation mimic well those of the natural environment. Cultivation efforts have been focused mainly on increasing dissolved CH$_4$ concentrations. To enrich AOM *ex situ*, batch and continuous reactors operated at moderate and high pressures have been tested. To avoid potential. sulfide toxicity, attention has been paid to exchange the medium so that the sulfide concentrations do not exceed 10 to 14 mM (Nauhaus et al., 2007; Nauhaus et al., 2002).

40

2.5.2 Conventional *in vitro* ANME enrichment techniques

The conventional *in vitro* incubation in gas tight serum bottles provides an opportunity to test the microbial activities, kinetics of the metabolic reactions and the enrichment of the microbes, more specifically for the large number of uncultured anaerobes like ANME. Conventional serum-bottles are widely used when the incubation pressures do not exceed 2.5 atm (Beal et al., 2009; Blumenberg et al., 2005; Holler et al., 2011a; Meulepas et al., 2009b).

A batch bottle experiment provides the flexibility to operate many different experiments in parallel (large numbers of experimental bottles can be handled at the same time) by controlling different environmental conditions such as temperature, salinity or alkalinity. The batch incubation based experiments are relatively easy to control and manipulate, especially with very slow growing microbes like ANME, which require strictly anaerobic conditions. AOM activity is negligible in the presence of oxygen (Treude et al., 2005b). The commonly used batch serum bottles or culture tubes with thick butyl rubber septa facilitate the sampling while maintaining the redox inside, although there are several other factors which can be key for ANME enrichment, such as the low solubility of CH_4 and the possible accumulation of sulfide toxicity in the stationary batches.

As shown in Table 2.1 and Table 2.3, several studies estimated the AOM rate by *in vitro* batch incubations (Holler et al., 2011b; Kruger et al., 2008; Wegener et al., 2008). Kruger et al. (2008) determined AOM rates from 4000 to 20000 nmol g_{dw}^{-1} d^{-1} by incubating microbial mats from the Black Sea. Holler et al. (2011a) estimated AOM at a rate of 250 nmol g_{dw}^{-1} d^{-1} by ANME-1 from the Black Sea (Table 2.1). ANME-2 dominated communities from the Hydrate Ridge of northeast Pacific exhibit 20 times higher specific AOM rates (20 mmol g_{dw}^{-1} d^{-1}) compared to ANME-1 from the Black Sea pink microbial mat (Nauhaus et al., 2005). During the *in vitro* incubations with different environmental conditions, unlike the SO_4^{2-} concentration, pH and salinity variations, temperature was found to be a major influential parameter for AOM rates in ANME-1 and ANME-2 communities (Nauhaus et al., 2005). Both ANME communities showed the increment in AOM rate with elevated CH_4 partial pressure. However, when the microbial mat from the Black Sea with both ANME-1 and ANME-2 was incubated in batch at low CH_4 concentrations, ANME-1 growth was favored over the growth of ANME-2 (Blumenberg et al., 2005).

Optimum pH, temperature, salinity and sulfide toxicity were determined as 7.5, 20°C, 30 ‰ and 2.5 mM, respectively, for the ANME-2 enrichment from Eckernförde Bay when incubated in 35 ml serum bottles (Meulepas et al., 2009b). The highest *in vitro* AOM activity was obtained at 15°C compared to other temperature incubations (Treude et al., 2005b) and sulfide toxicity was reported beyond 2.5 mM for Eckernförde Bay sediments (Meulepas et al., 2009b). Likewise, possible electron donors and acceptors involved in the AOM process were studied in the batch incubations. The sediment from Eckernförde Bay was incubated with different methanogenic substrates for the study of possible intermediates between the ANME and SRB (Meulepas et al., 2010a). The AOM activity with other electron acceptors than SO_4^{2-}, i.e., Fe (III) and Mn (IV), by Eel river sediment was estimated by batch incubations for the detection of Fe (III) / Mn (IV), dependent AOM (Beal et al., 2009). Moreover, thermophilic AOM was studied in batch assays within different temperature ranges (up to 100°C) with Guaymas Basin hydrothermal vent sediment, AOM was observed up to 75°C with the highest AOM rate at 50°C (Holler et al., 2011b).

2.5.3 Modified in vitro ANME enrichment approaches

The growth of ANME-2 was documented (Nauhaus et al., 2007) in batch incubations using a glass tube connected via a needle to a syringe and placed inside a pressure-proof steel cylinder (Nauhaus et al., 2002). The syringe, which is filled with medium, transmits the pressure of the cylinder to the medium inside the tube. Using this design, CH_4 hydrate sediment was incubated at 14 atm for 2 years with intermittent replenishment of the supernatant by fresh medium and CH_4 (21 mM at 12°C). During the incubation period, the volume of the ANME-2 and SRB consortia, which was tracked using FISH, increased exponentially (Nauhaus et al., 2007).

A batch incubation with intermittent replacement of supernatant by fresh medium (i.e., fed batch system) once a month was used to successfully enrich ANME-2d at abundances of about 78% (Haroon et al., 2013). The inoculum was a mixture of sediment from a local freshwater lake, anaerobic digester sludge and activated sludge from a wastewater treatment plant in Brisbane, Australia (Table 2.3) (Hu et al., 2009). The retention of biomass in the fed-batch system was achieved via a 20 min settling period prior to the replacement of the supernatant by fresh medium. The cultivation of this freshwater ANME-2d can have the advantage of higher solubility of CH_4 in freshwater than in seawater (Yamamoto et al., 1976), however, this microorganism was enriched at 35°C and CH_4 solubility decreases at increased temperatures (Hu et al., 2009). As previously specified (see section: AOM coupled to nitrite and nitrate reduction), this ANME-2d, named "*Candidatus Methanoperedens nitroreducens*", utilizes nitrate instead of SO_4^{2-} as electron acceptor for AOM. This physiological trait likely contributed to the successful enrichment of this novel ANME clade at high abundance in a relatively short time period (about 2 years), because AOM coupled to nitrate yields about 45-fold more energy than its SO_4^{2-} dependent counterpart (Figure 2.4).

2.5.4 Continuous bioreactor based ANME enrichment

The design rationale of a continuous flow incubation columns is to provide nutrients and to remove end products at environmentally relevant rates (Table 2.3) (Girguis et al., 2003). In such systems, 0.2 µm filtered seawater, reduced with hydrogen sulfide (510 µM) and saturated with CH_4 (1.5 mM) in a conditioning column (4 h at 5 atm), was used to feed cold seep and non-seep sediment cores maintained in PVC tubes at 2 atm and 5°C (Girguis et al., 2003). The AOM rates before and after incubations of the seep sediments were the same, probably because the incubation time was only 2 weeks. However, some increase in AOM rate and ANME-2c population size were detected in the non-seep sediment incubations. In a second experimental run, the same continuous flow reactor was used, but the incubations were conducted at 10 atm. The incubation time was 7.5 months (30 weeks) and a preferential proliferation of ANME-1 against ANME-2 was observed in the non-seep sediments at the highest pore water velocity tested (90 m year^{-1}) (Girguis et al., 2005). In addition, an increase in the AOM activity was reported as measured using batch incubations in serum bottles inoculated by the sediment (seep and non seep sediments used in the continuous enrichment experiment) without headspace, using 0.2 µm filter sterilized anoxic seawater containing 2 mM CH_4 and 1 mM hydrogen sulfide (Girguis et al., 2005).

Table 2.3 *Enrichment condition and AOM rate for the in vitro studies of AOM. Here, incubation temperature, pressure, ANME growth rate and apparent affinity is represented by T, P, μ and Km, respectively. DHS refers to downflow hanging sponge bioreactor. Different mineral medium used for incubation are indicated by the superscript letters next to each reference.* [a] *= the incubation in artificial salt water mineral medium prepared according to Widdel and Bak (1992),* [b] *= the incubation with filter sterilized sea water and* [c] *= incubation in fresh water medium with nitrate and ammonium. SR represents to* SO_4^{2-} *reduction.*

Enrichment mode	Innocula and incubation period (d)	T (°C)	P (bar)	ANME types	AOM rate ($\mu mol\ g_{dw}^{-1}\ d^{-1}$)	Doubling time of ANME (months)	μ (d^{-1})	Km (mM)	References
Fed-Batch	AOM and Annamox (230-290 d)	22-35	0.5-1	ANME-2d	1100 $\mu mol\ d^{-1}$	-	-	-	(Haroon et al., 2013)[c]
Membrane bioreactor, continuous well mixed	Baltic Sea/ Eckernförde Bay, 884 d	15	1	ANME-2a	286	3.8	0.006	< 0.5 mM (for SO_4^{2-}) 0.075 MPa (for CH_4)	(Meulepas et al., 2009a; Meulepas et al., 2009b)[a]
Fed-Batch	Hydrate Ridge, North-east Pacific, 700 d	15	14	ANME-2	230	7	0.003	> 10mM (CH_4)	(Nauhaus et al., 2007)[a]
Batch	Gulf of Mexico, 150 d	12	15	ANME-1	13.5	2	-	-	(Kruger et al., 2008)[a]
Fed-Batch	Gulf of Cadiz, 286 d	15	80	ANME-2	9.22 (SR)	2.5	-	37 mM (CH_4)	(Zhang et al., 2010; Zhang et al., 2011)[a]
Batch	Guaymas Basin sediment, 250 d	42-65	2.5	ANME-1	1.2	2.3	-	-	(Holler et al., 2011b)[a]
DHS bioreactor	Nankai Trough, 2013 d	10	1	ANME-1,-2,-3	0.375	-	-	-	(Aoki et al., 2014)[a]
AOM incubator system (continuous)	Monterey Bay, 400 d	5	1	ANME-1 ANME-2	9×10^{-3} (ANME-1) 0.138 (ANME-2)	1.1 (ANME-2) 1.4 (ANME-1)	0.03 (ANME-1) 0.024 (ANME-2)	-	(Girguis et al., 2005)[b]

In efforts to attain CH_4 concentrations close to *in situ* values, continuous reactors that can handle hydrostatic pressures up to 445 atm with CH_4 enriched medium and without free gas in the incubation chamber have been used (Deusner et al., 2009). This reactor configuration is flexible to operate in batch, fed-batch or continuous mode. Incubation of sediments from the Black Sea showed a six-fold increase in the volumetric AOM rate when the CH_4 partial pressure increased from 2 to 60 atm. In all operation modes, AOM rates were estimated based on sulfide production. However, when in otherwise similar operation conditions CH_4 saturated medium was replaced by CH_4 free medium, sulfide levels decreased rapidly and stabilized at input levels. This indicated that the sulfide production was indeed coupled to CH_4 oxidation. During continuous operation of such high pressure reactors, a CH_4 concentration of 60-65 mM can be readily attained. Noticeably, during continuous operation, the influent SO_4^{2-} concentration used was 8 mM, which is lower than seawater concentrations (Deusner et al., 2009). The hydraulic retention time was set at 14 h which corresponded to a dilution rate of 1.7 d^{-1}. Assuming a completely mixed reactor, this means that microorganisms growing at rates < 1.7 d^{-1} would be washed out from the reactor, which is the case of ANME having much lower growth rates (0.006 to 0.03 d^{-1}) (Girguis et al., 2005; Meulepas et al., 2009a; Meulepas et al., 2009b). Additionally, these tests of continuous operation with CH_4 addition lasted only 16 days and whether and how biomass was retained in the system was not reported (Deusner et al., 2009).

Similar high pressure systems have been operated at up to 600 atm hydrostatic pressure and 120°C (Sauer et al., 2012). The flexibility of this system allows the sub-sampling of medium without loss of pressure and it can be operated in batch or continuous mode (Sauer et al., 2012). The system was tested incubating sediments from the Isis mud volcano from the Egyptian continental margin (~ 991 m below sea level) using artificial seawater pre-conditioned with 40 atm of CH_4 resulting in dissolved concentrations of ~ 96 mM CH_4. Following CH_4 saturation, the hydrostatic pressure was increased to 100 atm using artificial seawater and incubations were conducted for 9 days at 23°C. No measurements of biomass concentration and yield were conducted, but an increase in sulfide was detected upon addition of CH_4 to the reactor (Sauer et al., 2012).

A continuous high pressure reactor capable to withstand up to 80 atm was used in fed-batch and continuous mode at pressures from 10 to 80 atm and a hydraulic retention time of 100 h during a 286 days incubation of sediments from a mud volcano located in the Gulf of Cadiz (Zhang et al., 2010). Under such conditions, the ANME-2 biovolume (count of cells and aggregates) increased about 12-fold (Zhang et al., 2011). There was no indication about the biomass retention time and AOM rate in the system.

ANME can also be enriched at moderate pressures or even ambient pressure provided biomass retention is applied. The latter can be achieved by introducing a submerged membrane (pore size 0.2 μm and effective surface of 0.03 m^2) within the reactor (Meulepas et al., 2009a). CH_4 was sparged continuously at 190 mmol L^{-1} d^{-1}, thus providing mixing, stripping-off of the sulfide and restricting fouling of the membrane. This bioreactor was operated at 15°C and at a slight over pressure (0.025 bars) to avoid oxygen intrusion. The SO_4^{2-} loading rate was 3 mmol L^{-1} d^{-1} and the hydraulic retention time 7 days. Sediment retrieved from the Eckernförde Bay in the Baltic Sea was used as inoculum and the reactor was operated for about 3 years. Growth of ANME was inferred by the increase in sulfide production in the membrane reactor, and the increase in AOM rates was monitored by carrying out batch experiment with reactor biomass amended with ^{13}C-labelled CH_4 at regular time intervals (Meulepas et al., 2009a). The ANMEs in the reactor could be affiliated to ANME-2a and their doubling time was estimated at 3.8 month (i.e., growth rate 0.006 d^{-1}).

Although high pressure reactors operate at high dissolved CH_4 concentrations, their maintenance and operation is cumbersome and requires meeting various safety criteria for their implementation. When successful enrichment has been reported at moderate pressures in fed-batch reactors, a key feature was a good biomass retention via settling (ANME-2d) (Haroon et al., 2013) or membranes (ANME-2a) (Meulepas et al., 2009a).

2.5.5 Future development in *ex situ* enrichment approaches

Mimicking the natural conditions in bioreactors can be a fruitful strategy for enrichment of ANME. Reproducing *in situ* conditions in the laboratory is quite challenging, but artificial material and equipment can be used to mimic the natural environment (Figure 2.9). Mimicking natural conditions is possible by using suitable reactors capable of achieving extreme environmental conditions such as high pressure or temperature and with suitable or similar natural packing material. The carbonate-minerals, where ANME have been found to form microbial reefs, are very porous. This porous natural matrix can harbor aggregates of AOM performing consortia (Marlow et al., 2014b). Similarly, polyurethane sponges are a porous material and can be used as packing material in a packed bed bioreactor configuration to promote the adhesion, aggregation and retention of biomass (Cassarini et al., 2017). The collected marine sediment can be entrapped in the porous sponges so that CH_4 can effectively diffuse through them, while the medium containing necessary nutrients and electron acceptor flows through the material (Imachi et al., 2011). In a recent study, fresh bituminous coal and sandstone collected from a coal mine were used in a flow through type reactor system at high pressure to simulate and study geological CO_2 sequestration and transformation (Ohtomo et al., 2013). Similarly, naturally occurring materials can be used as packing materials in bioreactors, which assist in biomass retention in the ANME enrichment bioreactor.

Considering the importance of substrate availability, especially for ANME which are oxidizing a poorly soluble compound like CH_4, membrane reactors can be used to facilitate the contact between substrate and biomass. A hollow-fiber membrane reactor was successfully applied for CH_4-dependent denitrification (Shi et al., 2013). CH_4 passes internally the hollow-fiber membranes and diffuses to the outside layer where a biofilm of ANME can be retained and grown (Figure 2.9). A silicone membrane can also be used as a hollow-fiber membrane, which allows bubbleless addition of gas to the bioreactor compartment. These gas diffusive membranes are also applicable for AOM-SR, where the diffused CH_4 can be immediately taken up by ANME consortia which are suspended in the SO_4^{2-} containing medium. This mode of CH_4 supply produces minimum bubbles and the gas supply can be controlled by maintaining the gas pressure inside the membrane. As the microbial metabolism of AOM is slow, the slow diffusion of CH_4 can reduce the large amounts of unused CH_4 released from a bioreactor system, thus reducing the operational costs. Another benefit of the membrane is the biomass retention, as the biomass usually develops as biofilm or flocs (Jagersma et al., 2009). Moreover, the sulfide and pH can be continuously monitored by using pH and pS (sulfide sensor) electrodes and the sulfide can be removed before reaching the toxic threshold. A process control algorithm has been developed for the SR process, which is also applicable for AOM studies (Cassidy et al., 2015) .

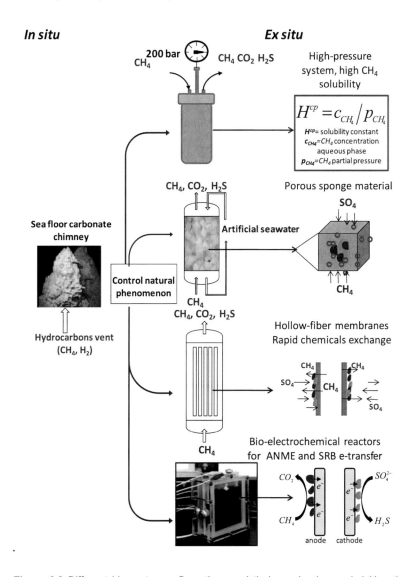

Figure 2.9 *Different bioreactor configurations and their mechanisms mimicking the growth mode of ANME in natural habitats to enhance ex situ growth of ANME.*

Several studies hypothesized an electron transfer between ANME and SRB (see section: *Syntrophy and potential electron transfer modes between ANME and SRB*). Based on this assumption, bio-electrochemical systems (BES) could also be used to enrich or isolate the ANME by replacing the SRB cells with a conductive membrane or electrode as electron sink. Thus, the CH_4 oxidation process by the ANME takes place at the anode and SR takes place at the cathode (Figure 2.9). The electron exchange between the electrodes and the ANME can be determined by applying different electrode potentials (Lovley, 2012). Another advantage of AOM studies using BES is the possibility to add compounds which can act as electron shuttles (e.g. electron mediators or conductive nanominerals such as iron oxides) between the electrodes and ANME to facilitate the electron transfer from the ANME cell to the electrode (Rabaey & Rozendal, 2010). Thus, the electrodes in BES facilitate experiments with electron transfer of CH_4 to the conducting surface and also serve as electron acceptor by which the ANME growth is possibly accelerated.

2.6 Approaches for AOM and ANME studies

2.6.1 Measurement of AOM rates in activity tests

Various geochemical and microbial analyses are carried out for ANME and AOM studies. The common approach used to identify the occurrence of AOM is by direct CH_4, CO_2, SO_4^{2-} and S^{2-} profile measurements in marine environments and batch incubations ((Reeburgh, 2007) *and reference therein*). However, measurement of the chemical profiles could not ensure whether the CO_2 and S^{2-} production is due to AOM or not. Therefore, other complementary methods such as *in vitro* incubation with stable isotopes or radioisotopes (e.g. $^{13}CH_4$ and $^{12}CH_4$) and profile measurement of labeled carbon are used for the estimation of AOM rate (by monitoring the $^{13}CH_4$ and $^{13}CO_2$ production in a batch) (Knittel & Boetius, 2009; Reeburgh, 2007). In addition, identification of the microbial community ensures the presence of ANME and establishes the link between the identity of the microorganisms and the AOM activity. A wide range of AOM rates have been observed in the different ANME habitats and bioreactors enrichments (Tables 2.1, 2.2 and 2.3).

2.6.2 ANME identification

Specific lipid biomarkers and stable carbon isotopes are often measured at potential AOM sites since the discovery of AOM (Blumenberg et al., 2005; Blumenberg et al., 2004; Hinrichs & Boetius, 2002; Hinrichs et al., 2000; Pancost et al., 2001; Rossel et al., 2008). Biomarkers are used to differentiate between archaeal and bacterial cells. Phospholipids fatty acids with an ether linkage are usually common for bacteria and eukarya (Niemann & Elvert, 2008). Distinction between ANME-1, ANME-2 and ANME-3 was explored by analysis of non-polar lipids and intact polar lipids as biomarkers (Rossel et al., 2008). ANME-1 contains a majority of isoprenoidal glycerol dialkyl glycerol tetraethers on its lipid profile, whereas ANME-2 and ANME-3 mostly contain phosphate-based polar derivatives of archaeol and hydroxyarchaeol (Niemann & Elvert, 2008; Rossel et al., 2008). While detection of lipid biomarkers provide information on the microorganism's identity, the carbon isotopic composition of the biomarkers provides information on the carbon source and/or metabolic fixation pathway of microbes (Hinrichs & Boetius, 2002).

CH_4 in marine environments is generally depleted in ^{13}C (carbon stable isotope composition, $\delta^{13}C$, of -50 to -110 ‰), while CO_2 is usually isotopically heavier than CH_4. Therefore, CH_4 oxidation would result in products which are depleted in ^{13}C. The finding of highly ^{13}C depleted lipids in archaeal biomass ($\delta^{13}C$ < -60‰) has been used as indicator of AOM with concomitant assimilation of carbon derived from the light CH_4 (^{12}C) by the ANME (Emerson & Hedges, 2008; Hinrichs & Boetius, 2002; Hinrichs et al., 1999; Hinrichs et al., 2000; Martens et al., 1999; Thomsen et al., 2001).

However, the carbon isotopic composition in many AOM habitats is complex. For instance, in cold seeps and vent sediments both the CH_4 and CO_2 are isotopically light. The isotopically light CO_2 is produced by chemoautotrophic microbes (Alperin & Hoehler, 2009), while the light CH_4 could be due to both methanogenesis and AOM (Pohlman et al., 2008). On the basis of these recent findings, the light isotope composition of the lipids in archaeal biomass indicated that the light carbon content can be assimilated via several processes (Figure 2.10): i) the assimilation from isotopically light CO_2 (^{13}C depleted CH_4 production rather than oxidation) by ANME, i.e., involvement of ANME-1 and ANME-2 in methanogenesis (Bertram et al., 2013; House et al., 2009), ii) the oxidation of CH_4 and utilization of inorganic carbon by ANME, i.e., autotrophic AOM by ANME-1 and ANME-2 (Kellermann et al., 2012), iii) AOM by the assimilation of ^{13}C depleted CH_4 in ANME, i.e., a common AOM process (Emerson & Hedges, 2008; Hinrichs & Boetius, 2002; Hinrichs et al., 1999; Hinrichs et al., 2000; Martens et al., 1999; Thomsen et al., 2001) and iv) ^{13}C depleted CO_2 assimilation by methanogens (Vigneron et al., 2015). Therefore, the conventional assumption of ^{13}C depleted archaeal biomarkers as a proxy for AOM has to be considered carefully. Moreover, the approach is not always straight forward for the depiction of AOM as light carbon in lipids can also originate from archaeal CH_4 production and not only from CH_4 oxidation (Alperin & Hoehler, 2009; Londry et al., 2008). Thus, the application of multiple approaches is advantageous for explicit understanding of AOM and ANME.

The confusion due to light lipid biomarkers from multiple carbon metabolisms can be partly overcome if stable isotope probing (SIP) is performed. ^{13}C enriched CH_4 and CO_2 can be used as substrates for *in vitro* incubations with the desired inoculum. Isotopic probing followed by lipids biomarker analysis (lipid-SIP) can be used to identify the carbon assimilation pathways for the microbes under investigation (Kellermann et al., 2012). Autotrophic and heterotrophic carbon assimilation together with lipid formation rates can be determined by dual lipid-SIP, which consists of simultaneous addition of deuterated water and ^{13}C-labeled inorganic carbon (Wegener et al., 2012). Moreover, the visualization of ANME cells or other molecular detection of ANME can be performed for clear elucidation on ANME occurrence.

Phylogenic analysis of 16s rRNA and *mcr*A genes from marine environments is generally performed for assigning identity to ANME types (Alain et al., 2006; Boetius et al., 2000; Harrison et al., 2009; Knittel et al., 2005; Losekann et al., 2007) (details described in the section *ANME phylogeny*). ANME cells are quantified by Q-PCR to assess ANME growth in enrichments and DNA finger print for comparison of ANME types among AOM sites (Girguis et al., 2005; Lloyd et al., 2011; Nunoura et al., 2008; Timmers et al., 2015; Wankel et al., 2012a). In the recent past, ANME specific primers were designed to enhance the quantification of particular ANME types (Miyashita et al., 2009; Zhou et al., 2014). Q-PCR can be performed to quantify the

RNA fraction of ANME genes, thus basically quantifying the active cells (Lloyd et al., 2010).

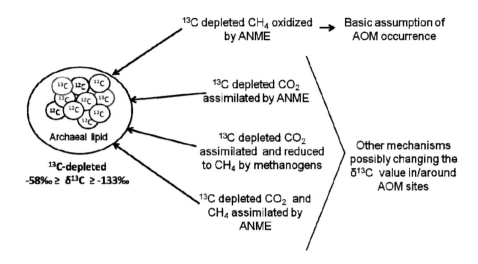

Figure 2.10 *Detection of ^{13}C depleted lipids in archaea as proxy of CH_4 oxidation: Basic assumption of AOM occurrence and other possible mechanisms that induce a change in the $\delta^{13}C$ value in archaeal lipids.*

Moreover, quantification of key functional genes such as *mcr*A (methanogenesis related) genes in ANME (Lee et al., 2013; Yanagawa et al., 2011) and *dsr*A (SO_4^{2-} reduction related) genes in SRB (Lee et al., 2013) were performed in recent studies. The analysis and quantification of specific functional genes allows the quantification of the microbes expressing the specific function only, so it will be easier to interpret the quantification results. The gene based analysis can nevertheless sometimes leads to a false conclusion. For example, the findings of specific DNA/RNA in a certain location may not always indicate the active cells in that location because cells might be transported from nearby active AOM areas. Hence, activity measurements of the biomass in those locations over time remain essential.

Recent studies on AOM pursued high throughput shot gun sequencing for the analysis of archaeal and bacterial communities in different sites including high temperature AOM (Mason et al., 2015; Wankel et al., 2012a). Small subunit ribosomal RNA (SSU rRNA) genes containing highly conserved and variable regions (V1-V9) were used as a marker for the high through put sequencing (Lynch & Neufeld, 2015). Among 23 distinct CH_4 seep sediments studied via pyrotag library analysis, ANME archaea and seep-SRB bacteria appeared as major communities in the cold anaerobic CH_4 seep, whereas aerobic methanotrophs and sulfide oxidizing *Thiotricales* groups were found mostly in the oxic part of CH_4 seeps (Ruff et al., 2015).

High throughput sequencing of specific gene amplicons provides information about the microbial community composition, whereas whole genomics analysis explores the functional profiles from gene to family level. Thus, the community metabolic pathways can be constructed on the basis of these genes (Franzosa et al., 2015). Chistoserdova et al. (2015) reviewed the aerobic and anaerobic methanotrophy on the basis of metagenomic studies. Metagenomics of ANME-1 and ANME-2a have been performed so far, supporting the reverse methanogenesis pathway for AOM-SR (Hallam et al., 2004; Meyerdierks et al., 2010). Moreover, the nitrate dependent AOM pathway was depicted by the ANME-2d genome in which the nitrate reductase specific gene was highly expressed (Haroon et al., 2013). Yet, more details on the omics based analysis of other ANME-phylotypes should be explored. Nevertheless, complete genomic studies are relevant for highly enriched ANME communities rather than the genomic analysis with sediment containing a few ANME cells for the explicit interpretation of genomic data and metabolic pathways. It should be noted that the genomic data provide mostly the information of the dominant community, so it is difficult to extract the information from the ANME genome if the amount of ANME genes is low in the sample analyzed.

2.6.3 ANME visualization and their functions studies

Fluorescence *in situ* hybridization (FISH) images provide insights regarding the morphology and the spatial arrangement of ANME and their bacterial associates (Blumenberg et al., 2004; Boetius et al., 2000; Knittel et al., 2005; Roalkvam et al., 2011). The FISH method has been widely discussed and applied in the past 25 years (Amann & Fuchs, 2008). CARD-FISH (catalyzed reported deposition–fluorescence *in situ* hybridization) with horseradish peroxidase (HRP) labeled probes are commonly used to visualize the ANME cells and SRBs in marine sediments (Holler et al., 2011b; Lloyd et al., 2011). The signal is amplified in CARD-FISH by using these probes together with fluorescently labeled tyramides, therefore copious fluorescent molecules can be introduced and the sensitivity increases compared to normal FISH

(Pernthaler et al., 2002). Detection of a few ANME cells by FISH does not always means that the studied site is an ANME habitat. For explicit AOM illustration, it is essential to combine FISH with other approaches such as activity measurements, quantification of ANME cells (by cell count, Q-PCR or quantification FISH), isotope probing or spectroscopic detection of metabolites.

In order to link the identity of microorganism to their functions, other methods that investigate the physiology and activities have to be combined with FISH. FISH in combination with SIMS (secondary ion mass spectroscopy) was used in AOM and ANME studies (Orphan et al., 2001b), which provides the linkage of ANME to its function by the visualization of ANME and analysis of compounds assimilated in the cells. The SIMS can analyze the isotopic composition of the cells so that it can be used to understand the mechanisms of AOM along with the syntrophy and intermediates (Orphan & House, 2009).

SIMS miniaturized instrumentation with a sub micrometer spatial resolution (NanoSIMS) has been used for ANME studies. It allows observation of single cell morphology in combination with FISH and quantitative analysis of the elemental and isotopic composition of cells with high sensitivity and precision (Behrens et al., 2008; Musat et al., 2008; Polerecky et al., 2012). FISH-NanoSIMS has been used in ANME studies detailing nitrogen fixation by ANME-2d archaea (Dekas et al., 2009) and sulfur metabolism in ANME-2 cells (Milucka et al., 2012). Normally highly enriched microbial communities are incubated with isotopic labeled substrates and the fate of the substrates is detected by specifically designed NanoSIMS equipment. FISH-NanoSIMS is often complemented by advanced microscopic observations such as scanning (SEM) or transmission (TEM) electron microscopy or atomic force microscopy (AFM) (Polerecky et al., 2012).

Microautoradiography-fluorescence *in situ* hybridization (MAR-FISH) is a promising approach to study ANME physiology by monitoring the assimilation of radio-labeled substrates by individual cells. The radio-labeled substrates (e.g. different carbon sources) are added to the samples containing active microbes and the fate of radioisotopes can be detected by MAR with simultaneous microbial identification by FISH (Lee et al., 1999; Nielsen & Nielsen, 2010). The handling of radioisotopes can limit the applications of this powerful technique.

Another appealing method for AOM studies is Raman spectroscopy combined with FISH (RAMAN-FISH) (Wagner, 2009), which analyzes the stable isotope at a micrometer level to provide the ecophysiology of a single cell. Raman microspectroscopy detects and quantifies the stable or radio isotope labeled substrate assimilation in the cell under study. The Raman spectra wavelength can provide distinction between the uptake of different substrates among cells exhibiting different metabolic pathways (Wagner, 2009). Raman spectroscopy can detect the molecular composition of a cell and thus can provide information on the molecules which are assimilated in the cells. The technique is thus useful for the study of assimilation of carbon and sulfur compounds in ANME. ANME cells can be visualized by FISH, and then examined with Raman spectroscopy for the substrate assimilation up to single cell level. It is highly applicable for the study of ANME cells as single cells from a complex microbial consortium can be analyzed and the assimilated compounds by these cells can be monitored.

2.6.4 New study approaches to AOM

ANME studies have immensely benefited from the advancement of microscopic and molecular tools. In recent years, several complementary approaches were applied for depicting AOM mechanisms, such as FISH-NanoSIMS together with Raman-FISH in a highly enriched ANME community (Milucka et al., 2012) and metagenomics together with FISH and continuous enrichment activity assays (Haroon et al., 2013). There are still several open questions to be addressed in ANME studies such as identification of intermediates, alternative substrates, tolerance limit for various environmental stresses and exploration of several ANME habitats. Also details of the carbon and sulfur metabolism by ANME and SRB are not elucidated to date.

Despite of advancement in genomic sequencing, the genome of only some phylotypes of ANME (ANME-2a,-2d and ANME-1) has been studied and the metabolic pathways were predicted. The predicted metabolic pathways by genomics can be verified by ecophysiological studies in combination with SIMS based spectroscopy. Further, many prospective approaches could be used for ANME studies to explore the ANME mechanism and ecophysiology, for example, an *in situ* SIP based survey for the study of AOM occurrence and carbon assimilation, single cell genomics for predicting metabolic pathways and genes from single cell isolates

(Rinke et al., 2014), imaging and mass spectroscopy of single cells or aggregates for understanding the metabolisms (Watrous & Dorrestein, 2011), and atomic force microscopy for the study of ANME cellular structure and detection of the effect of different stresses on the cell membrane (Dufrêne, 2014). Note that all these studies are only possible if the ANME microbial mats are appropriately handled from the seafloor to the laboratory or enriched in bioreactors.

In view of the complexity and lengthiness to cultivate a sufficient amount of ANME biomass, *in situ* investigations with a sophisticated *in situ* laboratory might overcome the current biomass handling and enrichment limitations. Taking advantage on the latest advances in deep sea instrumentation, which include various on-line data acquisition instruments, it should be possible for example to conduct *in situ* gas push-pull tests (i.e., tracer tests) (Urmann et al., 2004) which combined with *in situ* stable isotope probing approaches (Wankel et al., 2012b) and *in situ* molecular analysis can yield detailed microbial activity and function measurements in tandem with microbial identity. A deep sea environmental sample processor can be deployed up to 4000 m depth in ocean for sample collection, filtration and perform molecular analysis (Ussler III et al., 2013). It is capable to detect *in situ* and in real time molecular signals indicative of certain microorganisms or genes (Paul et al., 2007; Scholin et al., 2009). As proof-of-concept, an environmental sample processor has been used for the quantitative detection of 16S rRNA and particulate methane monooxygenase (*pmo*A) genes of aerobic methanotrophs near a CH_4-rich mound at a water depth of about 800 m (Ussler III et al., 2013). In principle, the environmental sample processor can be configured to detect and quantify genes and gene products from a wide range of microbial types (Preston et al., 2011). Additionally, the environmental sample processor is able to store samples for later *ex situ* validation analysis. The long term deployment capacity of the environmental sample processor is under development and this capability should allow temporal profiling of microorganisms which in tandem with on line characterization of physico-chemical parameters may help to understand which drivers are most important for the proliferation of active ANME communities in deep sea.

2.7 Conclusions and outlook

Undoubtedly much was learned about AOM in the last five decades, yet key knowledge gaps still exist. One of the most remarkable aspects requiring investigation relates to the proposed syntrophic association between ANME and SRB. Overall whether, when and how AOM occurs in obligatory syntrophic association with SRB remains unknown. If such syntrophy occurs, the intermediate compound serving as electron carrier between these two microorganisms or other mechanisms to shuttle the electrons remains unknown (Knittel & Boetius, 2009). AOM does not necessarily occur in a syntrophic association since ANME can oxidize CH_4 and reduce SO_4^{2-} to S^0 and potentially to sulfide (Milucka et al., 2012). The two canonical enzymes for dissimilatory SR were however not found in ANME cells, thus the enzymatic mechanisms of SR detected in ANME remains unknown (Milucka et al., 2012).

The marine habitats hosting ANME have been widely explored in the past, details on niche differentiation among the various ANME clades need to be further assessed. The presence and relevance of SO_4^{2-} dependent AOM in freshwater environments requires further exploration. Although a few investigations on freshwater habitats have been conducted, unambiguous links between the presence and activity of AOM are still required. Sediments.from eutrophic lakes and freshwater tidal creeks might be suitable locations to explore (Sivan et al., 2011). Yet, another aspect to resolve is the existence and identity of ANME directly utilizing iron or manganese oxides as electron acceptors in either marine or freshwater ecosystems. Some marine and brackish coastal locations having abundant iron oxides within CH_4 rich sediments have been identified. Sediments from those locations appear suitable for harboring ANME (Egger et al., 2015; Riedinger et al., 2014; Wankel et al., 2012a). Moreover, other naturally occurring electron acceptors such as selenate can be investigated in future AOM studies.

After a long effort, the most incommoding drawback is not being able to readily obtain enrichments of SO_4^{2-}dependent ANME. With the exception of a few studies in which ANME-2a enrichments were obtained after eight (Milucka et al., 2012) and three years (Meulepas et al., 2009a) in bioreactors, most biochemical studies have been conducted using naturally ANME enriched sediments of which the retrieved small quantities often limit experimental tests. In such context, proper handling of ANME biomass from the seafloor to the laboratory as well as the enrichment in bioreactor configurations mimicking *in situ* conditions are in priority. Alternatively, single cell microscopy and genomics as well as the development of an advanced *in situ* deep sea laboratory can help in unrevealing some of the remaining unknowns of the ANME metabolisms and ecophysiology.

2.8 References

Alain, K., Holler, T., Musat, F., Elvert, M., Treude, T., Krüger, M. 2006. Microbiological investigation of methane- and hydrocarbon- discharging mud volcanoes in the Carpathian Mountains, Romania. *Environ. Microbiol.*, **8**(4), 574-590.

Aller, R.C., Rude, P. 1988. Complete oxidation of solid phase sulfides by manganese and bacteria in anoxic marine sediments. *Geochim. Cosmochim. Ac.*, **52**(3), 751-765.

Alperin, M., Hoehler, T. 2010. The ongoing mystery of sea-floor methane. *Science*, **329**(5989), 288-289.

Alperin, M.J., Hoehler, T.M. 2009. Anaerobic methane oxidation by archaea/sulfate-reducing bacteria aggregates: 2. Isotopic constraints. *Am. J. Sci.*, **309**(10), 958-984.

Amann, R., Fuchs, B.M. 2008. Single-cell identification in microbial communities by improved fluorescence *in situ* hybridization techniques. *Nature. Rev. Microbiol.*, **6**(5), 339-348.

Amann, R.I., Ludwig, W., Schleifer, K.H. 1995. Phylogenetic identification and *in situ* detection of individual microbial cells without cultivation. *Microbiol. Rev.*, **59**(1), 143-69.

Aoki, M., Ehara, M., Saito, Y., Yoshioka, H., Miyazaki, M., Saito, Y., Miyashita, A., Kawakami, S., Yamaguchi, T., Ohashi, A., Nunoura, T., Takai, K., Imachi, H. 2014. A long-term cultivation of an anaerobic methane-oxidizing microbial community from deep-sea methane-seep sediment using a continuous-flow bioreactor. *PLoS One*, 9(8), Pe105356.

Archer, D., Buffett, B., Brovkin, V. 2009. Ocean methane hydrates as a slow tipping point in the global carbon cycle., *Proc. Natl. Acad. Sci. USA,* **106**(49), 20596-20601.

Beal, E.J., House, C.H., Orphan, V.J. 2009. Manganese- and iron-dependent marine methane oxidation. *Science*, **325**(5937), 184-7.

Behrens, S., Lösekann, T., Pett-Ridge, J., Weber, P.K., Ng, W.-O., Stevenson, B.S., Hutcheon, I.D., Relman, D.A., Spormann, A.M. 2008. Linking microbial phylogeny to metabolic activity at the single-cell level by using enhanced element labeling-catalyzed reporter deposition fluorescence in situ hybridization (EL-FISH) and NanoSIMS. *Appl. Environ. Microbiol.*, **74**(10), 3143-50.

Bertram, S., Blumenberg, M., Michaelis, W., Siegert, M., Krüger, M., Seifert, R. 2013. Methanogenic capabilities of ANME-archaea deduced from ^{13}C-labelling approaches. *Environ. Microbiol.*, **15**(8), 2384-2393.

Bhattarai, S., Cassarini, C., Gonzalez-Gil, G., Egger, M., Slomp, C.P., Zhang, Y., Esposito, G., and Lens, P.N. 2017. Anaerobic methane-oxidizing microbial community in a coastal marine sediment: anaerobic methanotrophy dominated by ANME-3. *Microb. Ecol.*, **74**(3), 608-622.

Biddle, J.F., Cardman, Z., Mendlovitz, H., Albert, D.B., Lloyd, K.G., Boetius, A., Teske, A. 2012. Anaerobic oxidation of methane at different temperature regimes in Guaymas Basin hydrothermal sediments. *ISME J.*, **6**(5), 1018-1031.

Blazewicz, S.J., Petersen, D.G., Waldrop, M.P., Firestone, M.K. 2012. Anaerobic oxidation of methane in tropical and boreal soils: Ecological significance in terrestrial methane cycling. *J. Geophys. Res..*, **117**(G2), 1-9.

Blumenberg, M., Seifert, R., Nauhaus, K., Pape, T., Michaelis, W. 2005. *In vitro* study of lipid biosynthesis in an anaerobically methane-oxidizing microbial mat. *Appl. Environ. Microbiol.*, **71**(8), 4345-4351.

Blumenberg, M., Seifert, R., Reitner, J., Pape, T., Michaelis, W. 2004. Membrane lipid patterns typify distinct anaerobic methanotrophic consortia. *Proc. Natl. Acad. Sci. USA,* **101**(30), 11111-11116.

Boetius, A., Ravenschlag, K., Schubert, C.J., Rickert, D., Widdel, F., Gieseke, A., Amann, R., Jorgensen, B.B., Witte, U., Pfannkuche, O. 2000. A marine microbial consortium apparently mediating anaerobic oxidation of methane. *Nature*, **407**(6804), 623-626.

Boetius, A., Wenzhöfer, F. 2013. Seafloor oxygen consumption fuelled by methane from cold seeps. *Nat. Geosci.*, **6**(9), 725-734.

Bradley, A.S., Fredricks, H., Hinrichs, K.-U., Summons, R.E. 2009. Structural diversity of diether lipids in carbonate chimneys at the Lost City hydrothermal field. *Org. Geochem.*, **40**(12), 1169-1178.

Brazelton, W.J., Schrenk, M.O., Kelley, D.S., Baross, J.A. 2006. Methane - and sulfur-metabolizing microbial communities dominate the Lost City hydrothermal field ecosystem. *Appl. Environ. Microbiol.*, **72**(9), 6257-6270.

Brooks, J.M., Field, M.E., Kennicutt, M.C. 1991. Observations of gas hydrates in marine sediments, offshore northern California. *Mar. Geol.*, **96**(1), 103-109.

Buffett, B., Archer, D. 2004. Global inventory of methane clathrate: sensitivity to changes in the deep ocean. *Earth Planet. Sci. Lett.*, **227**(3-4), 185-199.

Canfield, D.E., Thamdrup, B., Hansen, J.W. 1993. The anaerobic degradation of organic matter in Danish coastal sediments: iron reduction, manganese reduction, and sulfate reduction. *Geochim. Cosmochim. Ac.*, **57**(16), 3867-83.

Cassarini, C., Rene, E. R., Bhattarai, S., Esposito, G., Lens, P. N. 2017. Anaerobic oxidation of methane coupled to thiosulfate reduction in a biotrickling filter. *Biores. Technol.*, **240**, 214-222.

Cassidy, J., Lubberding, H.J., Esposito, G., Keesman, K.J., Lens, P.N. 2015. Automated biological sulphate reduction: a review on mathematical models, monitoring and bioprocess control. *FEMS Microbiol. Rev.*, **39**(6), 823-853.

Chistoserdova, L. 2015. Methylotrophs in natural habitats: current insights through metagenomics. *Appl. Environ. Microbiol.*, **99**(14), 5763-5779.

Chistoserdova, L., Vorholt, J.A., Lidstrom, M.E. 2005. A genomic view of methane oxidation by aerobic bacteria and anaerobic archaea. *Genome Biol.*, **6**(2), 1-6.

D'Hondt, S., Jørgensen, B.B., Miller, D.J., Batzke, A., Blake, R., Cragg, B.A., Cypionka, H., Dickens, G.R., Ferdelman, T., Hinrichs, K.-U., Holm, N.G., Mitterer, R., Spivack, A., Wang, G., Bekins, B., Engelen, B., Ford, K., Gettemy, G., Rutherford, S.D., Sass, H., Skilbeck, C.G., Aiello, I.W., Guèrin, G., House, C.H., Inagaki, F., Meister, P., Naehr, T., Niitsuma, S., Parkes, R.J., Schippers, A., Smith, D.C., Teske, A., Wiegel, J., Padilla, C.N., Acosta, J.L.S. 2004. Distributions of microbial activities in deep subseafloor sediments. *Science*, **306**(5705), 2216-2221.

D'Hondt, S., Rutherford, S., Spivack, A.J. 2002. Metabolic activity of subsurface life in deep-sea sediments. *Science*, **295**(5562), 2067-2070.

Dahl, C., Prange, A. 2006. Bacterial sulfur globules: occurrence, structure and metabolism. in: Shively, J. (Ed.), *Inclusions in Prokaryotes*. Vol. 1. Springer Berlin Heidelberg, Germany, pp. 21-51.

Dale, A.W., Van Cappellen, P., Aguilera, D.R., Regnier, P. 2008. Methane efflux from marine sediments in passive and active margins: estimations from bioenergetic reaction-transport simulations. *Earth Planet. Sci. Lett.*, **265**(3-4), 329-344.

Dekas, A.E., Chadwick, G.L., Bowles, M.W., Joye, S.B., Orphan, V.J. 2014. Spatial distribution of nitrogen fixation in methane seep sediment and the role of the ANME archaea. *Environ. Microbiol.*, **16**(10), 3012-3029.

Dekas, A.E., Poretsky, R.S., Orphan, V.J. 2009. Deep-Sea archaea fix and share nitrogen in methane-consuming microbial consortia. *Science*, **326**(5951), 422-426.

Deusner, C., Meyer, V., Ferdelman, T. 2009. High-pressure systems for gas-phase free continuous incubation of enriched marine microbial communities performing anaerobic oxidation of methane. *Biotechnol. Bioeng.*, **105**(3), 524-533.

Dufrêne, Y.F. 2014. Atomic force microscopy in microbiology: new structural and functional insights into the microbial cell surface. *MBio.*, 5(4), e01363.

Durisch-Kaiser, E., Klauser, L., Wehrli, B., Schubert, C. 2005. Evidence of intense archaeal and bacterial methanotrophic activity in the Black Sea water column. *Appl. Environ. Microbiol.*, **71**(12), 8099-8106.

Egger, M., Rasigraf, O., Sapart, C.I.J., Jilbert, T., Jetten, M.S., Röckmann, T., Van der Veen, C., Banda, N., Kartal, B., Ettwig, K.F., Slomp, C.P. 2015. Iron-mediated anaerobic oxidation of methane in brackish coastal sediments. *Environ. Sci.Technol.*, **49**(1), 277-283.

Emerson, S., Hedges, J. 2008. Stable and radioactive isotopes. in: *Chemical oceanography and the marine carbon cycle*, 1st Edition Cambridge University Press, pp. 468.

Ettwig, K.F., Butler, M.K., Le Paslier, D., Pelletier, E., Mangenot, S., Kuypers, M.M.M., Schreiber, F., Dutilh, B.E., Zedelius, J., de Beer, D., Gloerich, J., Wessels, H.J.C.T., van Alen, T., Luesken, F., Wu, M.L., van de Pas-Schoonen, K.T., Op den Camp, H.J.M., Janssen-Megens, E.M., Francoijs, K.-J., Stunnenberg, H., Weissenbach, J., Jetten, M.S.M., Strous, M. 2010. Nitrite-driven anaerobic methane oxidation by oxygenic bacteria. *Nature*, **464**(7288), 543-548.

Franzosa, E.A., Hsu, T., Sirota-Madi, A., Shafquat, A., Abu-Ali, G., Morgan, X.C., Huttenhower, C. 2015. Sequencing and beyond: integrating molecular 'omics' for microbial community profiling. *Nat. Rev. Microbiol.*, **13**(6), 360-372.

Gauthier, M., Bradley, R.L., Šimek, M. 2015. More evidence that anaerobic oxidation of methane is prevalent in soils: Is it time to upgrade our biogeochemical models? *Soil Biol. Biochem.*, **80**, 167-174.

Girguis, P.R., Cozen, A.E., DeLong, E.F. 2005. Growth and population dynamics of anaerobic methane-oxidizing archaea and sulfate-reducing bacteria in a continuous-flow bioreactor. *Appl. Environ. Microbiol.*, **71**(7), 3725-3733.

Girguis, P.R., Orphan, V.J., Hallam, S.J., DeLong, E.F. 2003. Growth and methane oxidation rates of anaerobic methanotrophic archaea in a continuous-flow bioreactor. *Appl. Environ. Microbiol.*, **69**(9), 5472-5482.

Gonzalez-Gil, G., Meulepas, R.J.W., Lens, P.N.L. 2011. Biotechnological aspects of the use of methane as electron donor for sulfate reduction. in: Murray, M.-Y.

(Ed.),*Comprehensive Biotechnology*. Vol. 6. (2nd edition), Elsevier B.V., Amsterdam, the Netherlands pp. 419-434.

Hallam, S.J., Girguis, P.R., Preston, C.M., Richardson, P.M., DeLong, E.F. 2003. Identification of methyl coenzyme M reductase A (*mcrA*) genes associated with methane-oxidizing archaea. *Appl. Environ. Microbiol*, **69**(9), 5483-5491.

Hallam, S.J., Putnam, N., Preston, C.M., Detter, J.C., Rokhsar, D., Richardson, P.M., DeLong, E.F. 2004. Reverse methanogenesis: testing the hypothesis with environmental genomics. *Science*, **305**(5689), 1457-1462.

Hanson, R.S., Hanson, T.E. 1996. Methanotrophic bacteria. *Microbiol. Rev.*, **60**(2), 439-471.

Harder, J. 1997. Anaerobic methane oxidation by bacteria employing ^{14}C-methane uncontaminated with ^{14}C-carbon monoxide. *Mar. Geol.*, **137**(1-2), 13-23.

Haroon, M.F., Hu, S., Shi, Y., Imelfort, M., Keller, J., Hugenholtz, P., Yuan, Z., Tyson, G.W. 2013. Anaerobic oxidation of methane coupled to nitrate reduction in a novel archaeal lineage. *Nature*, **500**(7464), 567-570.

Harrison, B.K., Zhang, H., Berelson, W., Orphan, V.J. 2009. Variations in archaeal and bacterial diversity associated with the sulfate-methane transition zone in continental margin sediments (Santa Barbara Basin, California). *Appl. Environ. Microbiol.*, **75**(6), 1487-1499.

Haynes, C.A., Gonzalez, R. 2014. Rethinking biological activation of methane and conversion to liquid fuels. *Nat. Chem. Biol.*, **10**(5), 331-339.

Heijs, S.K., Haese, R.R., van der Wielen, P.W., Forney, L.J., van Elsas, J.D. 2007. Use of 16S rRNA gene based clone libraries to assess microbial communities potentially involved in anaerobic methane oxidation in a Mediterranean cold seep. *Microb. Ecol.*, **53**(3), 384-398.

Hinrichs, K.-U., Boetius, A. 2002. The Anaerobic Oxidation of Methane: New Insights in Microbial Ecology and Biogeochemistry. in: Wefer, G., Billett, D., Hebbeln, D., Jørgensen, B., Schlüter, M., van Weering, T.E. (Eds.), *Ocean Margin Systems*. Springer Berlin Heidelberg, Germany, pp. 457-477.

Hinrichs, K.-U., Hayes, J.M., Sylva, S.P., Brewer, P.G., DeLong, E.F. 1999. Methane-consuming archaebacteria in marine sediments. *Nature*, **398**(6730), 802-805.

Hinrichs, K.-U., Summons, R.E., Orphan, V.J., Sylva, S.P., Hayes, J.M. 2000. Molecular and isotopic analysis of anaerobic methane-oxidizing communities in marine sediments. *Org. Geochem.*, **31**(12), 1685-1701.

Hoehler, T.M., Alperin, M.J., Albert, D.B., Martens, S. 1994. Field and laboratory studies of methane oxidation in an anoxic marine sediment: Evidence for a methanogen-sulfate reducer consortium. *Global Biogeochem. Cy.*, **8**(4), 451-463.

Holler, T., Wegener, G., Knittel, K., Boetius, A., Brunner, B., Kuypers, M., M. M., Widdel, F. 2009. Substantial $^{13}C/^{12}C$ and D/H fractionation during anaerobic oxidation of methane by marine consortia enriched *in vitro*. *Environ. Microbiol. Rep.*, **1**(5), 370-376.

Holler, T., Wegener, G., Niemann, H., Deusner, C., Ferdelman, T.G., Boetius, A., Brunner, B., Widdel, F. 2011a. Carbon and sulfur back flux during anaerobic microbial oxidation of methane and coupled sulfate reduction. *Proc. Nat. Acad. Sci. USA.*, **108**(52), E1484-E1490.

Holler, T., Widdel, F., Knittel, K., Amann, R., Kellermann, M.Y., Hinrichs, K.-U., Teske, A., Boetius, A., Wegener, G. 2011b. Thermophilic anaerobic oxidation of methane by marine microbial consortia. *ISME J.*, **5**(12), 1946-1956.

House, C.H., Orphan, V.J., Turk, K., A., Thomas, B., Pernthaler, A., Vrentas, J.M., Joye, S., B. . 2009. Extensive carbon isotopic heterogeneity among methane seep microbiota. *Environ. Microbiol.*, **11**(9), 2207-2215.

Hu, S., Zeng, R.J., Burow, L.C., Lant, P., Keller, J., Yuan, Z. 2009. Enrichment of denitrifying anaerobic methane oxidizing microorganisms. *Environ. Microbiol. Rep.*, **1**(5), 377-384.

Imachi, H., Aoi, K., Tasumi, E., Saito, Y.Y., Yamanaka, Y., Yamaguchi, T., Tomaru, H., Takeuchi, R., Morono, Y., Inagaki, F., Takai, K. 2011. Cultivation of methanogenic community from subseafloor sediments using a continuous-flow bioreactor. *ISME J.*, **5**(12), 1913-25.

IPCC. 2007. Climate change 2007: the physical science basis. in: Solomon, S., Qin, D., Manning, M., Chen, Z., Marquis, M., Averyt, K.B., Tignor M., Miller, H.L. (Eds.), *Contribution of working group I to the fourth assessment report of the intergovernmental panel on climate change*, Cambridge University Press, New York, USA., pp. 996.

Iversen, N., Jorgensen, B.B. 1985. Anaerobic methane oxidation rates at the sulfate-methane transition in marine sediments from Kattegat and Skagerrak (Denmark). *Limnol. Oceanogr.*, **30**(5), 944-955.

Jagersma, G.C., Meulepas, R.J.W., Heikamp-de Jong, I., Gieteling, J., Klimiuk, A., Schouten, S., Sinninghe Damsté, J.S., Lens, P.N., Stams, A.J. 2009. Microbial diversity and community structure of a highly active anaerobic methane-oxidizing sulfate-reducing enrichment. *Environ. Microbiol.*, **11**(12), 3223-3232.

Jørgensen, B., Kasten, S. 2006. Sulfur Cycling and Methane Oxidation. in: Schulz, H., Zabel, M. (Eds.), *Marine Geochemistry*. (2nd edition), Springer Berlin Heidelberg, pp. 271-309.

Joye, S.B., Boetius, A., Orcutt, B.N., Montoya, J.P., Schulz, H.N., Erickson, M.J., Lugo, S.K. 2004. The anaerobic oxidation of methane and sulfate reduction in sediments from Gulf of Mexico cold seeps. *Chem. Geol.*, **205**(3), 219-238.

Kato, S., Hashimoto, K., Watanabe, K. 2012a. Methanogenesis facilitated by electric syntrophy via (semi)conductive iron-oxide minerals. *Environ. Microbiol*, **14**(7), 1646-1654.

Kato, S., Hashimoto, K., Watanabe, K. 2012b. Microbial interspecies electron transfer via electric currents through conductive minerals. *Proc. Nat. Acad. Sci. USA.*, **109**(25), 10042-10046.

Kellermann, M.Y., Wegener, G., Elvert, M., Yoshinaga, M.Y., Lin, Y.-S., Holler, T., Mollar, X.P., Knittel, K., Hinrichs, K.-U. 2012. Autotrophy as a predominant mode of carbon fixation in anaerobic methane-oxidizing microbial communities. *Proc. Nat. Acad. Sci. USA.*, **109**(47), 19321-19326.

Kirschke, S., Bousquet, P., Ciais, P., Saunois, M., Canadell, J.G., Dlugokencky, E.J., Bergamaschi, P., Bergmann, D., Blake, D.R., Bruhwiler, L., Cameron-Smith, P., Castaldi, S., Chevallier, F., Feng, L., Fraser, A., Heimann, M., Hodson, E.L., Houweling, S., Josse, B., Fraser, P.J., Krummel, P.B., Lamarque, J.-F., Langenfelds, R.L., Le Quere, C., Naik, V., O'Doherty, S., Palmer, P.I., Pison,

I., Plummer, D., Poulter, B., Prinn, R.G., Rigby, M., Ringeval, B., Santini, M., Schmidt, M., Shindell, D.T., Simpson, I.J., Spahni, R., Steele, L.P., Strode, S.A., Sudo, K., Szopa, S., van der Werf, G.R., Voulgarakis, A., van Weele, M., Weiss, R.F., Williams, J.E., Zeng, G. 2013. Three decades of global methane sources and sinks. *Nat. Geosci.*, **6**(10), 813-823.

Knittel, K., Boetius, A. 2009. Anaerobic oxidation of methane: progress with an unknown process. *Annu. Rev. Microbiol.*, **63**(1), 311-334.

Knittel, K., Lösekann, T., Boetius, A., Kort, R., Amann, R. 2005. Diversity and distribution of methanotrophic archaea at cold seeps. *Appl. Environ. Microbiol.*, **71**(1), 467-479.

Kormas, K., Meziti, A., Dählmann, A., De Lange, G., Lykousis, V. 2008. Characterization of methanogenic and prokaryotic assemblages based on mcrA and 16S rRNA gene diversity in sediments of the Kazan mud volcano (Mediterranean Sea). *Geobiology*, **6**(5), 450-460.

Krause, S., Steeb, P., Hensen, C., Liebetrau, V., Dale, A.W., Nuzzo, M., Treude, T. 2014. Microbial activity and carbonate isotope signatures as a tool for identification of spatial differences in methane advection: a case study at the Pacific Costa Rican margin. *Biogeosciences*, **11**, 507-523.

Krüger, M., Blumenberg, M., Kasten, S., Wieland, A., Känel, L., Klock, J.-H., Michaelis, W., Seifert, R. 2008. A novel, multi-layered methanotrophic microbial mat system growing on the sediment of the Black Sea. *Environ. Microbiol.*, **10**(8), 1934-1947.

Krüger, M., Meyerdierks, A., Glockner, F.O., Amann, R., Widdel, F., Kube, M., Reinhardt, R., Kahnt, R., Bocher, R., Thauer, R.K., Shima, S. 2003. A conspicuous nickel protein in microbial mats that oxidize methane anaerobically. *Nature*, **426**(6968), 878-881.

Krüger, M., Wolters, H., Gehre, M., Joye, S.B., Richnow, H.-H. 2008. Tracing the slow growth of anaerobic methane-oxidizing communities by ^{15}N-labelling techniques. *FEMS Microbiol. Ecol.*, **63**(3), 401-411.

Lanoil, B.D., Sassen, R., La Duc, M.T., Sweet, S.T., Nealson, K.H. 2001. *Bacteria* and *Archaea* physically associated with Gulf of Mexico gas hydrates. *Appl. Environ. Microbiol.*, **67**(11), 5143-5153.

Larowe, D.E., Dale, A.W., Regnier, P. 2008. A thermodynamic analysis of the anaerobic oxidation of methane in marine sediments. *Geobiology*, **6**(5), 436-449.

Lazar, C.S., John Parkes, R., Cragg, B.A., L'Haridon, S., Toffin, L. 2012. Methanogenic activity and diversity in the centre of the Amsterdam Mud Volcano, Eastern Mediterranean Sea. *FEMS Microbiol. Ecol.*, **81**(1), 243-254.

Lee, J.-W., Kwon, K.K., Azizi, A., Oh, H.-M., Kim, W., Bahk, J.-J., Lee, D.-H., Lee, J.-H. 2013. Microbial community structures of methane hydrate-bearing sediments in the Ulleung Basin, East Sea of Korea. *Mar. Petrol. Geol.*, **47**, 136-146.

Lee, N., Nielsen, P.H., Andreasen, K.H., Juretschko, S., Nielsen, J.L., Schleifer, K.-H., Wagner, M. 1999. Combination of fluorescent *in situ* hybridization and microautoradiography-a new tool for structure-function analyses in microbial ecology. *Appl. Environ. Microbiol.*, **65**(3), 1289-1297.

Lens, P., Vallero, M., Esposito, G., Zandvoort, M.H. 2002. Perspectives of sulfate reducing bioreactors in environmental biotechnology. *Rev. Environ. Sci. Biotechnol.*, **1**(4), 311-325.

Lever, M.A., Rouxel, O., Alt, J.C., Shimizu, N., Ono, S., Coggon, R.M., Shanks, W.C., Lapham, L., Elvert, M., Prieto-Mollar, X., Hinrichs, K.-U., Inagaki, F., Teske, A. 2013. Evidence for microbial carbon and sulfur cycling in deeply buried ridge flank basalt. *Science*, **339**(6125), 1305-1308.

Lloyd, K.G., Alperin, M.J., Teske, A. 2011. Environmental evidence for net methane production and oxidation in putative ANaerobic MEthanotrophic (ANME) archaea. *Environ. Microbiol.*, **13**(9), 2548-2564.

Lloyd, K.G., Lapham, L., Teske, A. 2006. An anaerobic methane-oxidizing community of ANME-1b archaea in hypersaline Gulf of Mexico sediments. *Appl. Environ. Microbiol.*, **72**(11), 7218-7230.

Lloyd, K.G., MacGregor, B.J., Teske, A. 2010. Quantitative PCR methods for RNA and DNA in marine sediments: maximizing yield while overcoming inhibition. *FEMS Microbiol. Ecol.*, **72**(1), 143-151.

Londry, K.L., Dawson, K.G., Grover, H.D., Summons, R.E., Bradley, A.S. 2008. Stable carbon isotope fractionation between substrates and products of *Methanosarcina barkeri*. *Org. Geochem.*, **39**(5), 608-621.

Losekann, T., Knittel, K., Nadalig, T., Fuchs, B., Niemann, H., Boetius, A., Amann, R. 2007. Diversity and abundance of aerobic and anaerobic methane oxidizers at the Haakon Mosby mud volcano, Barents Sea. *Appl. Environ. Microbiol.*, **73**(10), 3348-3362.

Lovley, D.R. 2012. Electromicrobiology. *Annu. Rev. Microbiology*, **66**(1), 391-409.

Lovley, D.R. 2008. Extracellular electron transfer: wires, capacitors, iron lungs, and more. *Geobiology*, **6**(3), 225-231.

Lynch, M.D., Neufeld, J.D. 2015. Ecology and exploration of the rare biosphere. *Nat. Rev. Microbiol.*, **13**(4), 217-229.

Maignien, L., Parkes, R.J., Cragg, B., Niemann, H., Knittel, K., Coulon, S., Akhmetzhanov, A., Boon, N. 2013. Anaerobic oxidation of methane in hypersaline cold seep sediments. *FEMS Microbiol. Ecol.*, **83**(1), 214-231.

Marlow, J.J., Steele, J.A., Case, D.H., Connon, S.A., Levin, L.A., Orphan, V.J. 2014a. Microbial abundance and diversity patterns associated with sediments and carbonates from the methane seep environments of Hydrate Ridge, OR. *Front. Mar. Sci.*, **1**(44), 1-16.

Marlow, J.J., Steele, J.A., Ziebis, W., Thurber, A.R., Levin, L.A., Orphan, V.J. 2014b. Carbonate-hosted methanotrophy represents an unrecognized methane sink in the deep sea. *Nat. Commun.*, **5**(5094), 1-12.

Martens, C.S., Albert, D.B., Alperin, M. 1999. Stable isotope tracing of anaerobic methane oxidation in the gassy sediments of Eckernforde Bay, German Baltic Sea. *A. J. Sci.*, **299**(7-9), 589-610.

Martens, C.S., Berner, R.A. 1974. Methane production in the interstitial waters of sulfate-depleted marine sediments. *Science*, **185**(4157), 1167-1169.

Mason, O.U., Case, D.H., Naehr, T.H., Lee, R.W., Thomas, R.B., Bailey, J.V., Orphan, V.J. 2015. Comparison of archaeal and bacterial diversity in methane seep carbonate nodules and host sediments, Eel River Basin and Hydrate Ridge, USA. *Microbial. Ecol.*, 70(3), 766-784.

Mayr, S., Latkoczy, C., Krüger, M., Günther, D., Shima, S., Thauer, R.K., Widdel, F., Jaun, B. 2008. Structure of an F430 variant from archaea associated with anaerobic oxidation of methane. *J. Am. Chem. Soc.*, **130**(32), 10758-10767.

Merkel, A.Y., Huber, J.A., Chernyh, N.A., Bonch-Osmolovskaya, E.A., Lebedinsky, A.V. 2013. Detection of putatively thermophilic anaerobic methanotrophs in diffuse hydrothermal vent fluids. *Appl. Environ. Microbiol.*, **79**(3), 915-923.

Meulepas, R.J.W., Jagersma, C.G., Gieteling, J., Buisman, C.J.N., Stams, A.J.M., Lens, P.N.L. 2009a. Enrichment of anaerobic methanotrophs in sulfate-reducing membrane bioreactors. *Biotechnol. Bioeng.*, **104**(3), 458-470.

Meulepas, R.J.W., Jagersma, C.G., Khadem, A., Stams, A.J.W., Lens, P.N.L. 2010a. Effect of methanogenic substrates on anaerobic oxidation of methane and sulfate reduction by an anaerobic methanotrophic enrichment. *Appl. Microbiol. Biotechnol.*, **87**(4), 1499-1506.

Meulepas, R.J.W., Jagersma, C.G., Khadem, A.F., Buisman, C.J.N., Stams, A.J.M., Lens, P.N.L. 2009b. Effect of environmental conditions on sulfate reduction with methane as electron donor by an Eckernförde Bay enrichment. *Environ. Sci. Technol.*, **43**(17), 6553-6559.

Meulepas, R.J.W., Jagersma, C.G., Zhang, Y., Petrillo, M., Cai, H., Buisman, C.J.N., Stams, A.J.W., Lens, P.N.L. 2010b. Trace methane oxidation and the methane dependency of sulfate reduction in anaerobic granular sludge. *FEMS Microbiol. Ecol.*, **72**(2), 261-271.

Meulepas, R.J.W., Stams, A.J.M., Lens, P.N.L. 2010c. Biotechnological aspects of sulfate reduction with methane as electron donor. *Rev. Environ. Sci. Biotechnol.*, **9**(1), 59-78.

Meyerdierks, A., Kube, M., Kostadinov, I., Teeling, H., Glöckner, F.O., Reinhardt, R., Amann, R. 2010. Metagenome and mRNA expression analyses of anaerobic methanotrophic archaea of the ANME-1 group. *Environ. Microbiol.*, **12**(2), 422-439.

Michaelis, W., Seifert, R., Nauhaus, K., Treude, T., Thiel, V., Blumenberg, M., Knittel, K., Gieseke, A., Peterknecht, K., Pape, T., Boetius, A., Amann, R., Jørgensen, B.B., Widdel, F., Peckmann, J., Pimenov, N.V., Gulin, M.B. 2002. Microbial reefs in the Black Sea fueled by anaerobic oxidation of methane. *Science*, **297**(5583), 1013-1015.

Milucka, J., Ferdelman, T.G., Polerecky, L., Franzke, D., Wegener, G., Schmid, M., Lieberwirth, I., Wagner, M., Widdel, F., Kuypers, M.M.M. 2012. Zero-valent sulphur is a key intermediate in marine methane oxidation. *Nature*, **491**(7425), 541-546.

Miyashita, A., Mochimaru, H., Kazama, H., Ohashi, A., Yamaguchi, T., Nunoura, T., Horikoshi, K., Takai, K., Imachi, H. 2009. Development of 16S rRNA gene-targeted primers for detection of archaeal anaerobic methanotrophs (ANMEs). *FEMS Microbiol. Lett.*, **297**(1), 31-37.

Moran, J.J., Beal, E.J., Vrentas, J.M., Orphan, V.J., Freeman, K.H., House, C.H. 2008. Methyl sulfides as intermediates in the anaerobic oxidation of methane. *Environ. Microbiol.*, **10**(1), 162-173.

Musat, N., Halm, H., Winterholler, B., Hoppe, P., Peduzzi, S., Hillion, F., Horreard, F., Amann, R., Jørgensen, B.B., Kuypers, M.M.M. 2008. A single-cell view on the ecophysiology of anaerobic phototrophic bacteria. *Proc. Nat. Acad. Sci. USA.*, **105**(46), 17861-6.

Muyzer, G., Stams, A.J. 2008. The ecology and biotechnology of sulphate-reducing bacteria. *Nat. Rev. Microbiol.*, **6**(6), 441-454.

Nauhaus, K., Albrecht, M., Elvert, M., Boetius, A., Widdel, F. 2007. *In vitro* cell growth of marine archaeal-bacterial consortia during anaerobic oxidation of methane with sulfate. *Environ. Microbiol.*, **9**(1), 187-196.

Nauhaus, K., Boetius, A., Kruger, M., Widdel, F. 2002. *In vitro* demonstration of anaerobic oxidation of methane coupled to sulphate reduction in sediment from a marine gas hydrate area. *Environ. Microbiol.*, **4**(5), 296-305.

Nauhaus, K., Treude, T., Boetius, A., Kruger, M. 2005. Environmental regulation of the anaerobic oxidation of methane: a comparison of ANME-I and ANME-II communities. *Environ. Microbiol.*, 7(1), 98-106.

Nazaries, L., Murrell, J.C., Millard, P., Baggs, L., Singh, B.K. 2013. Methane, microbes and models: fundamental understanding of the soil methane cycle for future predictions. *Environ. Microbiol.*, **15**(9), 2395-2417.

Nielsen, J., Nielsen, P. 2010. Combined microautoradiography and fluorescence in situ hybridization (MAR-FISH) for the identification of metabolically active microorganisms. in: Timmis, K.N. (Ed.), *Handbook of Hydrocarbon and Lipid Microbiology*, Vol 1. Springer Berlin Heidelberg, pp. 4093-4102.

Nielsen, L.P., Risgaard-Petersen, N., Fossing, H., Christensen, P.B., Sayama, M. 2010. Electric currents couple spatially separated biogeochemical processes in marine sediment. *Nature*, **463**(7284), 1071-1074.

Niemann, H., Duarte, J., Hensen, C., Omoregie, E., Magalhaes, V.H., Elvert, M., Pinheiro, L.M., Kopf, A., Boetius, A. 2006a. Microbial methane turnover at mud volcanoes of the Gulf of Cadiz. *Geochim. Cosmochim. Ac.*, **70**(21), 5336-5355.

Niemann, H., Elvert, M. 2008. Diagnostic lipid biomarker and stable carbon isotope signatures of microbial communities mediating the anaerobic oxidation of methane with sulphate. *Org. Geochem.*, **39**(12), 1668-1677.

Niemann, H., Elvert, M., Hovland, M., Orcutt, B., Judd, A., Suck, I., Gutt, J., Joye, S., Damm, E., Finster, K., Boetius, A. 2005. Methane emission and consumption at a North Sea gas seep (Tommeliten area). *Biogeosciences*, **2**, 335-351.

Niemann, H., Losekann, T., de Beer, D., Elvert, M., Nadalig, T., Knittel, K., Amann, R., Sauter, E.J., Schluter, M., Klages, M., Foucher, J.P., Boetius, A. 2006b. Novel microbial communities of the Haakon Mosby mud volcano and their role as a methane sink. *Nature*, **443**(7113), 854-858.

Novikova, S.A., Shnyukov, Y.F., Sokol, E.V., Kozmenko, O.A., Semenova, D.V., Kutny, V.A. 2015. A methane-derived carbonate build-up at a cold seep on the Crimean slope, north-western Black Sea. *Mar. Geol.*, **363**, 160-173.

Nunoura, T., Oida, H., Miyazaki, J., Miyashita, A., Imachi, H., Takai, K. 2008. Quantification of mcrA by fluorescent PCR in methanogenic and methanotrophic microbial communities. *FEMS Microbiol. Ecol.*, **64**(2), 240-247.

Ohtomo, Y., Ijiri, A., Ikegawa, Y., Tsutsumi, M., Imachi, H., Uramoto, G.-I., Hoshino, T., Morono, Y., Sakai, S., Saito, Y., Tanikawa, W., Hirose, T., Inagaki, F. 2013. Biological CO_2 conversion to acetate in subsurface coal-sand formation using a high-pressure reactor system. *Front. Microbiol.*, **4**(361), 1-17.

Oni, O. E., Miyatake, T., Kasten, S., Richter-Heitmann, T., Fischer, D., Wagenknecht, L., Ksenofontov, V., Kulkarni, A., Blumers, M., Shylin, S. 2015. Distinct microbial populations are tightly linked to the profile of dissolved iron in the methanic sediments of the Helgoland mud area, North Sea. *Front. Microbiol.* **6**(365), 1-15.

Omoregie, E.O., Niemann, H., Mastalerz, V., Lange, G.J.D., Stadnitskaia, A., Mascle, J., Foucher, J.P., Boetius, A. 2009. Microbial methane oxidation and sulfate reduction at cold seeps of the deep Eastern Mediterranean Sea. *Mar. Geol.*, **261**(1-4), 114-127.

Orcutt, B., Boetius, A., Elvert, M., Samarkin, V., Joye, S.B. 2005. Molecular biogeochemistry of sulfate reduction, methanogenesis and the anaerobic oxidation of methane at Gulf of Mexico cold seeps. *Geochim. Cosmochim. Ac.*, **69**(17), 4267-4281.

Orcutt, B., Samarkin, V., Boetius, A., Joye, S. 2008. On the relationship between methane production and oxidation by anaerobic methanotrophic communities from cold seeps of the Gulf of Mexico. *Environ. Microbiol.*, **10**(5), 1108-1117.

Orphan, V.J., Hinrichs, K.-U., Ussler, W., Paull, C.K., Taylor, L.T., Sylva, S.P., Hayes, J.M., Delong, E.F. 2001a. Comparative analysis of methane-oxidizing archaea and sulfate-reducing bacteria in anoxic marine sediments. *Appl. Environ. Microbiol.*, **67**(4), 1922-1934.

Orphan, V.J., House, C.H. 2009. Geobiological investigations using secondary ion mass spectrometry: microanalysis of extant and paleo-microbial processes. *Geobiology*, **7**(3), 360-372.

Orphan, V.J., House, C.H., Hinrichs, K.-U., McKeegan, K.D., DeLong, E.F. 2001b. Methane-consuming archaea revealed by directly coupled isotopic and phylogenetic analysis. *Science*, **293**(5529), 484-487.

Orphan, V.J., House, C.H., Hinrichs, K.-U., McKeegan, K.D., DeLong, E.F. 2002. Multiple archaeal groups mediate methane oxidation in anoxic cold seep sediments. *Proc. Nat. Acad. Sci. USA.*, **99**(11), 7663-7668.

Orphan, V.J., Ussler, W., Naehr, T.H., House, C.H., Hinrichs, K.-U., Paull, C.K. 2004. Geological, geochemical, and microbiological heterogeneity of the seafloor around methane vents in the Eel River Basin, offshore California. *Chem. Geol.*, **205**(3-4), 265-289.

Pachiadaki, M., G. , Lykousis, V., Stefanou, E., G., Kormas, K., A. 2010. Prokaryotic community structure and diversity in the sediments of an active submarine mud volcano (Kazan mud volcano, East Mediterranean Sea). *FEMS Microbiol. Ecol.*, **72**(3), 429-444.

Pachiadaki, M.G., Kallionaki, A., Dählmann, A., De Lange, G.J., Kormas, K.A. 2011. Diversity and spatial distribution of prokaryotic communities along a sediment vertical profile of a deep-sea mud volcano. *Microb. Ecol.*, **62**(3), 655-668.

Pancost, R.D., Hopmans, E.C., Damste, J.S.S. 2001. Archaeal lipids in Mediterranean cold seeps: Molecular proxies for anaerobic methane oxidation. *Geochim. Cosmochim. Ac.*, **65**(10), 1611-1627.

Parkes, R.J., Cragg, B.A., Banning, N., Brock, F., Webster, G., Fry, J.C., Hornibrook, E., Pancost, R.D., Kelly, S., Knab, N., Jørgensen, B.B., Rinna, J., Weightman, A.J. 2007. Biogeochemistry and biodiversity of methane cycling in subsurface marine sediments (Skagerrak, Denmark). *Environ. Microbiol.*, **9**(5), 1146-1161.

Parkes, R.J., Cragg, B.A., Wellsbury, P. 2000. Recent studies on bacterial populations and processes in subseafloor sediments: A review. *Hydrogeol. J.*, **8**(1), 11-28.

Paul, J., SCholin, C., Van Den Engh, G., Perry, M.J. 2007. *In situ* instrumentation. *Oceanography*, **20**(2), 70-78.

Pernthaler, A., Pernthaler, J., Amann, R. 2002. Fluorescence *in situ* hybridization and catalyzed reporter deposition for the identification of marine bacteria. *Appl. Environ. Microbiol.*, **68**(6), 3094-3101.

Pinero, E., Marquardt, M., Hensen, C., Haeckel, M., Wallmann, K. 2013. Estimation of the global inventory of methane hydrates in marine sediments using transfer functions. *Biogeosciences*, **10**(2), 959-975.

Pohlman, J.W., Ruppel, C., Hutchinson, D.R., Downer, R., Coffin, R.B. 2008. Assessing sulfate reduction and methane cycling in a high salinity pore water system in the northern Gulf of Mexico. *Mar. Petrol. Geol.*, **25**(9), 942-951.

Polerecky, L., Adam, B., Milucka, J., Musat, N., Vagner, T., Kuypers, M.M. 2012. Look@ NanoSIMS–a tool for the analysis of nanoSIMS data in environmental microbiology. *Environ. Microbiol.*, **14**(4), 1009-1023.

Preston, C.M., Harris, A., Ryan, J.P., Roman, B., Marin III, R., Jensen, S., Everlove, C., Birch, J., Dzenitis, J.M., Pargett, D. 2011. Underwater application of quantitative PCR on an ocean mooring. *PLoS One*, **6**(8), e22522.

Pruesse, E., Peplies, J., Glöckner, F.O. 2012. SINA: accurate high-throughput multiple sequence alignment of ribosomal RNA genes. *Bioinformatics*, **28**(14), 1823-1829.

Rabaey, K., Rozendal, R.A. 2010. Microbial electrosynthesis-revisiting the electrical route for microbial production. *Nat. Rev. Microbiol.*, **8**(10), 706-716.

Raghoebarsing, A.A., Pol, A., van de Pas-Schoonen, K.T., Smolders, A.J.P., Ettwig, K.F., Rijpstra, W.I.C., Schouten, S., Damsté, J.S.S., Op den Camp, H.J.M., Jetten, M.S.M., Strous, M. 2006. A microbial consortium couples anaerobic methane oxidation to denitrification. *Nature*, **440**(7086), 918-21.

Reeburgh, W.S. 1980. Anaerobic methane oxidation: rate depth distributions in Skan Bay sediments. *Earth Planet. Sci. Lett.*, **47**(3), 345-352.

Reeburgh, W.S. 1976. Methane consumption in Cariaco Trench waters and sediments. *Earth Planet. Sci. Lett.*, **28**(3), 337-344.

Reeburgh, W.S. 2007. Oceanic methane biogeochemistry. *Chem. Rev.*, **107**(2), 486-513.

Reguera, G., McCarthy, K.D., Mehta, T., Nicoll, J.S., Tuominen, M.T., Lovley, D.R. 2005. Extracellular electron transfer via microbial nanowires. *Nature*, **435**(7045), 1098-1101.

Reitner, J., Peckmann, J., Reimer, A., Schumann, G., Thiel, V. 2005. Methane-derived carbonate build-ups and associated microbial communities at cold seeps on the lower Crimean shelf (Black Sea). *Facies*, **51**(1-4), 66-79.

Riedinger, N., Formolo, M.J., Lyons, T.W., Henkel, S., Beck, A., Kasten, S. 2014. An inorganic geochemical argument for coupled anaerobic oxidation of methane and iron reduction in marine sediments. *Geobiology*, **12**(2), 172-181.

Rinke, C., Lee, J., Nath, N., Goudeau, D., Thompson, B., Poulton, N., Dmitrieff, E., Malmstrom, R., Stepanauskas, R., Woyke, T. 2014. Obtaining genomes from uncultivated environmental microorganisms using FACS–based single-cell genomics. *Nat. Protoc.*, **9**(5), 1038-1048.

Roalkvam, I., Dahle, H., Chen, Y., Jørgensen, S.L., Haflidason, H., Steen, I.H. 2012. Fine-scale community structure analysis of ANME in Nyegga sediments with high and low methane flux. *Front. Microbiol.*, **3**(216), 1-13.

Roalkvam, I., Jørgensen, S.L., Chen, Y., Stokke, R., Dahle, H., Hocking, W.P., Lanzén, A., Haflidason, H., Steen, I.H. 2011. New insight into stratification of anaerobic methanotrophs in cold seep sediments. *FEMS Microbiol. Ecol.*, **78**(2), 233-243.

Rossel, P.E., Lipp, J.S., Fredricks, H.F., Arnds, J., Boetius, A., Elvert, M., Hinrichs, K.-U. 2008. Intact polar lipids of anaerobic methanotrophic archaea and associated bacteria. *Org. Geochem.*, **39**(8), 992-999.

Ruff, S.E., Biddle, J.F., Teske, A.P., Knittel, K., Boetius, A., Ramette, A. 2015. Global dispersion and local diversification of the methane seep microbiome. *Proc. Nat. Acad. Sci. USA.*, **112**(13), 4015-4020.

Saitou, N., Nei, M. 1987. The neighbor-joining method: a new method for reconstructing phylogenetic trees. *Mol. Biol. Evol.*, **4**(4), 406-425.

Sauer, P., Glombitza, C., Kallmeyer, J. 2012. A system for incubations at high gas partial pressure. *Front. Microbiol.*, **3**(25), 225-233.

Scheller, S., Goenrich, M., Boecher, R., Thauer, R.K., Jaun, B. 2010. The key nickel enzyme of methanogenesis catalyses the anaerobic oxidation of methane. *Nature*, **465**(7298), 606-608.

Scholin, C., Doucette, G., Jensen, S., Roman, B., Pargett, D., Marin III, R., Preston, C., Jones, W., Feldman, J., Everlove, C. 2009. Remote detection of marine microbes, small invertebrates, harmful algae, and biotoxins using the environmental sample processor (ESP). *Oceanography*, **22**, 158-167.

Schreiber, L., Holler, T., Knittel, K., Meyerdierks, A., Amann, R. 2010. Identification of the dominant sulfate-reducing bacterial partner of anaerobic methanotrophs of the ANME-2 clade. *Environ. Microbiol.*, **12**(8), 2327-2340.

Schubert, C.J., Coolen, M.J.L., Neretin, L.N., Schippers, A., Abbas, B., Durisch-Kaiser, E., Wehrli, B., Hopmans, E.C., Damste, J.S.S., Wakeham, S., Kuypers, M.M.M. 2006. Aerobic and anaerobic methanotrophs in the Black Sea water column. *Environ. Microbiol.*, **8**(10), 1844-1856.

Segarra, K.E.A., Comerford, C., Slaughter, J., Joye, S.B. 2013. Impact of electron acceptor availability on the anaerobic oxidation of methane in coastal freshwater and brackish wetland sediments. *Geochim. Cosmochim. Ac.*, **115**, 15-30.

Segarra, K.E.A., Schubotz, F., Samarkin, V., Yoshinaga, M.Y., Hinrichs, K.-U., Joye, S.B. 2015. High rates of anaerobic methane oxidation in freshwater wetlands reduce potential atmospheric methane emissions. *Nat. Commun.*, **6**(7477), 1-8.

Shi, Y., Hu, S., Lou, J., Lu, P., Keller, J., Yuan, Z. 2013. Nitrogen removal from wastewater by coupling anammox and methane-dependent denitrification in a membrane biofilm reactor. *Environ. Sci. Technol.*, **47**(20), 11577-11583.

Shima, S., Krueger, M., Weinert, T., Demmer, U., Kahnt, J., Thauer, R.K., Ermler, U. 2012. Structure of a methyl-coenzyme M reductase from Black Sea mats that oxidize methane anaerobically. *Nature*, **481**(7379), 98-101.

Shima, S., Thauer, R.K. 2005. Methyl-coenzyme M reductase and the anaerobic oxidation of methane in methanotrophic Archaea. *Curr. Opin. Microbiol.*, **8**(6), 643-648.

Sievert, S.M., Kiene, R.P., Schultz-Vogt, H.N. 2007. The sulfur cycle. *Oceanography*, **20**, 117-123.

Sivan, O., Adler, M., Pearson, A., Gelman, F., Bar-Or, I., John, S.G., Eckert, W. 2011. Geochemical evidence for iron-mediated anaerobic oxidation of methane. *Limnol. Oceanogr.*, **56**(4), 1536-1544.

Sivan, O., Antler, G., Turchyn, A.V., Marlow, J.J., Orphan, V.J. 2014. Iron oxides stimulate sulfate-driven anaerobic methane oxidation in seeps. *Proc. Nat. Acad. Sci. USA.*, **111**(40), E4139-E4147.

Sivan, O., Schrag, D.P., Murray, R.W. 2007. Rates of methanogenesis and methanotrophy in deep-sea sediments. *Geobiology*, **5**(2), 141-151.

Slomp, C.P., Mort, H.P., Jilbert, T., Reed, D.C., Gustafsson, B.G., Wolthers, M. 2013. Coupled dynamics of iron and phosphorus in sediments of an oligotrophic coastal basin and the impact of anaerobic oxidation of methane. *PLoS One*, **8**(6), e62386.

Sørensen, K., Finster, K., Ramsing, N. 2001. Thermodynamic and kinetic requirements in anaerobic methane oxidizing consortia exclude hydrogen, acetate, and methanol as possible electron shuttles. *Microb. Ecol.*, **42**(1), 1-10.

Stadnitskaia, A., Muyzer, G., Abbas, B., Coolen, M.J.L., Hopmans, E.C., Baas, M., van Weering, T.C.E., Ivanov, M.K., Poludetkina, E., Sinninghe Damste, J.S. 2005. Biomarker and 16S rDNA evidence for anaerobic oxidation of methane and related carbonate precipitation in deep-sea mud volcanoes of the Sorokin Trough, Black Sea. *Mar. Geol.*, **217**(1-2), 67-96.

Stams, A.J., Plugge, C.M. 2009. Electron transfer in syntrophic communities of anaerobic bacteria and archaea. *Nat. Rev. Microbiol.*, **7**(8), 568-577.

Straub, K.L., Schink, B. 2004. Ferrihydrite-dependent growth of *Sulfurospirillum deleyianum* through electron transfer via sulfur cycling. *Appl. Environ. Microbiol.*, **70**(10), 5744-5749.

Suess, E. 2014. Marine cold seeps and their manifestations: geological control, biogeochemical criteria and environmental conditions. *Int. J. Earth Sci.*, **103**(7), 1889-1916.

Tavormina, P.L., Ussler, W., Joye, S.B., Harrison, B.K., Orphan, V.J. 2010. Distributions of putative aerobic methanotrophs in diverse pelagic marine environments. *ISME J.*, **4**(5), 700-710.

Teske, A., Hinrichs, K.-U., Edgcomb, V., Gomez, A.D., Kysela, D., Sylva, S.P., Sogin, M.L., Jannasch, H.W. 2002. Microbial diversity of hydrothermal sediments in the Guaymas Basin: Evidence for anaerobic methanotrophic communities. *Appl. Environ. Microbiol.*, **68**(4), 1994-2007.

Thauer, R. K., Shima, S. 2008. Methane as fuel for anaerobic microorganisms. *Ann. N. Y. Acad. Sci.*, **1125**(1), 158-170.

Thauer, R. K. 2011. Anaerobic oxidation of methane with sulfate: on the reversibility of the reactions that are catalyzed by enzymes also involved in methanogenesis from CO_2. *Curr. Opin. Microbiol.*, **14**(3), 292-299.

Thiel, V., Peckmann, J., Richnow, H.H., Luth, U., Reitner, J., Michaelis, W. 2001. Molecular signals for anaerobic methane oxidation in Black Sea seep carbonates and a microbial mat. *Mar. Chem.*, **73**(2), 97-112.

Thomsen, T.R., Finster, K., Ramsing, N.B. 2001. Biogeochemical and molecular signatures of anaerobic methane oxidation in a marine sediment *Appl. Environ. Microbiol.*, **67**(4), 1646-1656.

Timmers, P.H., Gieteling, J., Widjaja-Greefkes, H.A., Plugge, C.M., Stams, A.J., Lens, P.N., Meulepas, R.J. 2015. Growth of anaerobic methane-oxidizing archaea and sulfate-reducing bacteria in a high-pressure membrane capsule bioreactor. *Appl. Environ. Microbiol.*, **81**(4), 1286-1296.

Treude, T., Knittel, K., Blumenberg, M., Seifert, R., Boetius, A. 2005a. Subsurface microbial methanotrophic mats in the Black Sea. *Appl. Environ. Microbiol.*, **71**(10), 6375-6378.

Treude, T., Krüger, M., Boetius, A., Jørgensen, B.B. 2005b. Environmental control on anaerobic oxidation of methane in the gassy sediments of Eckernförde Bay (German Baltic). *Limnol. Oceanogr.*, **50**(6), 1771-1786.

Treude, T., Orphan, V.J., Knittel, K., Gieseke, A., House, C.H., Boetius, A. 2007. Consumption of methane and CO_2 by methanotrophic microbial mats from gas seeps of the anoxic Black Sea. *Appl. Environ. Microbiol.*, **73**(7), 2271-2283.

Urmann, K., Gonzalez-Gil, G., Schroth, M.H., Hofer, M., Zeyer, J. 2004. New field method: gas push-pull test for the in-situ quantification of microbial activities in the vadose zone. *Environ. Sci. Technol.*, **39**(1), 304-310.

Ussler III, W., Preston, C., Tavormina, P., Pargett, D., Jensen, S., Roman, B., Marin III, R., Shah, S.R., Girguis, P.R., Birch, J.M. 2013. Autonomous application of quantitative PCR in the deep sea: in situ surveys of aerobic methanotrophs using the deep-sea environmental sample processor. *Environ. Sci. Technol.*, **47**(16), 9339-9346.

Valentine, D.L. 2002. Biogeochemistry and microbial ecology of methane oxidation in anoxic environments: a review. *Anton. Leeuw. Int. J. G.*, **81**(1-4), 271-282.

Valentine, D.L., Reeburgh, W.S., Hall, R. 2000. New perspectives on anaerobic methane oxidation. *Environ. Microbiol.*, **2**(5), 477-484.

Vasquez-Cardenas, D., van de Vossenberg, J., Polerecky, L., Malkin, S.Y., Schauer, R., Hidalgo-Martinez, S., Confurius, V., Middelburg, J.J., Meysman, F.J.R., Boschker, H.T.S. 2015. Microbial carbon metabolism associated with electrogenic sulphur oxidation in coastal sediments. *ISME J.*, **9**(9), 1966-1978.

Vigneron, A., Cruaud, P., Pignet, P., Caprais, J.-C., Cambon-Bonavita, M.-A., Godfroy, A., Toffin, L. 2013. Archaeal and anaerobic methane oxidizer communities in the Sonora Margin cold seeps, Guaymas Basin (Gulf of California). *ISME J.*, **7**(8), 1595-1608.

Vigneron, A., L'Haridon, S., Godfroy, A., Roussel, E.G., Cragg, B.A., Parkes, R.J., Toffin, L. 2015. Evidence of active methanogen communities in shallow sediments of the Sonora Margin cold seeps. *Appl. Environ. Microbiol.*, **81**(10), 3451-3459.

Wagner, M. 2009. Single-cell ccophysiology of microbes as revealed by Raman Microspectroscopy or Secondary Ion Mass Spectrometry Imaging. *Ann. Rev. Microbiol.*, **63**(1), 411-429.

Wallmann, K., Pinero, E., Burwicz, E., Haeckel, M., Hensen, C., Dale, A., Ruepke, L. 2012. The global inventory of methane hydrate in marine sediments: A theoretical approach. *Energies*, **5**(7), 2449-2498.

Wan, M., Shchukarev, A., Lohmayer, R., Planer-Friedrich, B., Peiffer, S. 2014. Occurrence of surface polysulfides during the interaction between ferric (hydr)oxides and aqueous sulfide. *Environ. Sci. Technol.*, **48**(9), 5076-5084.

Wang, F.-P., Zhang, Y., Chen, Y., He, Y., Qi, J., Hinrichs, K.-U., Zhang, X.-X., Xiao, X., Boon, N. 2014. Methanotrophic archaea possessing diverging methane-oxidizing and electron-transporting pathways. *ISME J.*, **8**(5), 1069-1078.

Wankel, S.D., Adams, M.M., Johnston, D.T., Hansel, C.M., Joye, S.B., Girguis, P.R. 2012a. Anaerobic methane oxidation in metalliferous hydrothermal sediments: influence on carbon flux and decoupling from sulfate reduction. *Environ. Microbiol.*, **14**(10), 2726-2740.

Wankel, S.D., Huang, Y.-w., Gupta, M., Provencal, R., Leen, J.B., Fahrland, A., Vidoudez, C., Girguis, P.R. 2012b. Characterizing the distribution of methane sources and cycling in the deep sea via in situ stable isotope analysis. *Environ. Sci. Technol.*, **47**(3), 1478-1486.

Wankel, S.D., Joye, S.B., Samarkin, V.A., Shah, S.R., Friederich, G., Melas-Kyriazi, J., Girguis, P.R. 2010. New constraints on methane fluxes and rates of anaerobic methane oxidation in a Gulf of Mexico brine pool via *in situ* mass spectrometry. *Deep Sea Res. Part II: Top. Stud. Oceanogr.*, **57**(21-23), 2022-2029.

Watrous, J.D., Dorrestein, P.C. 2011. Imaging mass spectrometry in microbiology. *Nat. Rev. Microbiol.*, **9**(9), 683-694.

Wegener, G., Bausch, M., Holler, T., Thang, N.M., Prieto Mollar, X., Kellermann, M.Y., Hinrichs, K.-U., Boetius, A. 2012. Assessing sub-seafloor microbial activity by combined stable isotope probing with deuterated water and ^{13}C-bicarbonate. *Environ. Microbiol.*, **14**(6), 1517-1527.

Wegener, G., Niemann, H., Elvert, M., Hinrichs, K.-U., Boetius, A. 2008. Assimilation of methane and inorganic carbon by microbial communities mediating the anaerobic oxidation of methane. *Environ. Microbiol.*, **10**(9), 2287-2298.

Widdel, F., Bak, F. 1992. Gram negative mesophilic sulfate reducing bacteria. in: Balows, A., Truper, H., Dworkin, M., Harder, W., Schleifer, K.H. (Eds.), *The* *prokaryotes: a handbook on the biology of bacteria: ecophysiology, isolation, identification, applications,*Vol. II. Springer New York, USA, pp. 3352-3378.

Widdel, F., Hansen, T. 1992. The dissimilatory sulfate-and sulfur-reducing bacteria. in: Balows, A., Truper, H., Dworkin, M., Harder, W., Schleifer, K.H. (Eds.) *The prokaryotes: a handbook on the biology of bacteria: ecophysiology, isolation, identification, applications.* Vol. I (2nd edition), Springer New York, USA, pp. 582-624.

Wrede, C., Brady, S., Rockstroh, S., Dreier, A., Kokoschka, S., Heinzelmann, S.M., Heller, C., Reitner, J., Taviani, M., Daniel, R., Hoppert, M. 2012. Aerobic and anaerobic methane oxidation in terrestrial mud volcanoes in the Northern Apennines. *Sediment. Geol.*, **263-264**, 210-219.

Yamamoto, S., Alcauskas, J.B., Crozier, T.E. 1976. Solubility of methane in distilled water and seawater. *J. Chem. Eng. Data*, **21**(1), 78-80.

Yanagawa, K., Sunamura, M., Lever, M.A., Morono, Y., Hiruta, A., Ishizaki, O., Matsumoto, R., Urabe, T., Inagaki, F. 2011. Niche separation of methanotrophic archaea (ANME-1 and-2) in methane-seep sediments of the eastern Japan Sea offshore Joetsu. *Geomicrobiol. J.*, **28**(2), 118-129.

Zehnder, A.J., Brock, T.D. 1980. Anaerobic methane oxidation: occurrence and ecology. *Appl. Environ. Microbiol.*, **39**(1), 194-204.

Zehnder, A.J.B., Brock, T.D. 1979. Methane formation and methane oxidation by methanogenic bacteria. *J. Bacteriol.*, **137**, 420-432.

Zhang, Y., Henriet, J.-P., Bursens, J., Boon, N. 2010. Stimulation of in vitro anaerobic oxidation of methane rate in a continuous high-pressure bioreactor. *Bioresour. Technol.*, **101**(9), 3132-3138.

Zhang, Y., Maignien, L., Zhao, X., Wang, F., Boon, N. 2011. Enrichment of a microbial community performing anaerobic oxidation of methane in a continuous high-pressure bioreactor. *BMC Microbiol.*, **11**(137), 1-8.

Zhou, Z., Han, P., Gu, J.-D. 2014. New PCR primers based on *mcrA* gene for retrieving more anaerobic methanotrophic archaea from coastal reedbed sediments. *Appl. Microbiol. Biotechnol.*, **98**(10), 4663-4670.

CHAPTER 3

Microbial Sulfate Reducing Activities in Anoxic Sediment from Marine Lake Grevelingen

The modified version of this chapter was published as:

Bhattarai S., Cassarini C., Naangmenyele Z., Rene E. R,, Gonzalez-Gil G., Esposito G.and Lens P.N.L. (2017) Microbial sulfate reducing activities in anoxic sediment from Marine Lake Grevelingen: screening of electron donors and acceptors. *Limnology,* DOI: 10.1007/s10201-017-0516-0

Abstract

Sulfate reducing bacteria in marine sediment mainly utilize sulfate as a terminal electron acceptor with different organic compounds as electron donor. This study investigated microbial sulfate reducing activity of coastal sediment from the Marine Lake Grevelingen (the Netherlands) using different electron donors and electron acceptors. All four electron donors (ethanol, lactate, acetate and methane) showed sulfate reducing activity with sulfate as electron acceptor, suggesting the presence of an active sulfate reducing bacterial population in the sediment, even at dissolved sulfide concentrations exceeding 12 mM. Ethanol showed the highest sulfate reduction rate of 55 µmol g_{VSS}^{-1} d^{-1} compared to lactate (32 µmol g_{VSS}^{-1} d^{-1}), acetate (26 µmol g_{VSS}^{-1} d^{-1}) and methane (4.7 µmol g_{VSS}^{-1} d^{-1}). Sulfate reduction using thiosulfate and elemental sulfur as electron acceptors and methane as the electron donor was observed, however, mainly by disproportionation rather than by anaerobic oxidation of methane coupled to sulfate reduction. This study showed that the Marine Lake Grevelingen sediment is capable to perform sulfate reduction by using diverse electron donors, including the gaseous and cheap electron donor methane.

3.1 Introduction

Microbial sulfate reduction (SR) to sulfide is a ubiquitous process in marine sediments, where it is mainly fueled by microbial degradation of organic matter (Arndt et al., 2013) and/or by anaerobic oxidation of methane (AOM) (Knittel and Boetius 2009). This redox reaction is mediated by sulfate reducing bacteria (SRB) (Muyzer and Stams 2008), although they are widely distributed and play an active role in the sulfur cycle, the majority of SRB in marine sediments is unculturable and their physiology is thus poorly described (D'Hondt et al., 2004; Xiong et al., 2013).

The microbial SR process has been successfully applied in the industry for the biological treatment of wastewater containing sulfate (SO_4^{2-}) or other sulfur oxyanions such as thiosulfate, sulfite or dithionite, wherein the end product sulfide can be precipitated as elemental sulfur (S^0) after an aerobic post-treatment or as metal sulfides in case of metal containing wastewaters (Liamleam and Annachhatre 2007; Weijma et al., 2006). Often the necessity of additional electron donors, such as ethanol or hydrogen, for the SR is expensive; therefore, it is appealing to study the

84

activity and SR rates of SRB from diverse habitats and their performance using easily accessible and low-priced electron donors, such as methane (CH_4) (Gonzalez-Gil et al., 2011; Meulepas et al., 2010).

Table 3.1 Reactions and standard Gibb's free energy changes at pH 7.0 ($\Delta G^{0\prime}$) for methane, thiosulfate, elemental sulfur, ethanol, lactate and acetate during anaerobic sulfate reduction

Electron donor	Reaction	$\Delta G^{0\prime}$ [a] kJ mol^{-1} electron donor
Methane	$CH_4 + SO_4^{2-} \rightarrow HCO_3^- + HS^- + H_2O$	-17
	$CH_4 + S_2O_3^{2-} \rightarrow HCO_3^- + 2HS^- + H^+$	-39
	$CH_4 + 4S^0 + 3H_2O \rightarrow HCO_3^- + 4HS^- + 5H^+$	+24
Thiosulfate	$S_2O_3^{2-} + H_2O \rightarrow SO_4^{2-} + HS^- + H^+$	-22
Elemental sulfur	$4S^0 + 4H_2O \rightarrow SO_4^{2-} + 3HS^- + 5H^+$	+40
Ethanol	$2CH_3CH_2OH + SO_4^{2-}$ $\rightarrow 2CH_3COO^- + HS^- + H^+$ $+ 2H_2O$	-32
Lactate	$2CH_3CHOHCOO^- + SO_4^{2-}$ $\rightarrow 2CH_3COO^- + 2HCO_3^- + HS^-$ $+ H^+$	-38
	$3CH_3CHOHCOO^-$ $\rightarrow CH_3COO^- + 2C_2H_5COO^-$ $+ HCO_3^- + H^+$	-169
Acetate	$CH_3COO^- + SO_4^{2-} \rightarrow HS^- + 2HCO_3^-$	- 47

Note:

[a] The $\Delta G^{0\prime}$ values were calculated from Gibbs free energies of formation from the elements at standard temperature and pressure, as obtained from Thauer et al. (1977)

The main challenge of using AOM coupled to SR (AOM-SR) as a process for the desulfurization of wastewater is the slow growth rate of the microorganisms involved (Deusner et al., 2009; Krüger et al., 2008; Meulepas et al., 2009a; Nauhaus et al., 2007; Zhang et al., 2010), which could possibly be increased by using more thermodynamically favorable sulfur compounds other than SO_4^{2-}, such as thiosulfate ($S_2O_3^{2-}$)(Table 3.1) or S^0 which was reported to be an intermediate in the AOM induced SR (Milucka et al., 2012).

The coastal marine sediment from the Marine Lake Grevelingen (MLG), the Netherlands, has a special microbial ecology as it harbors both cable bacteria

(Hagens et al., 2015; Vasquez-Cardenas et al., 2015; Sulu-Gambari et al., 2016) and anaerobic methanotrophs (ANME) (Bhattarai et al., 2017). A recent study on geochemical data modelling has predicted that SR and methanogenesis might be prominent microbial processes in the MLG lake sediment, while a large amount of CH_4 could be diffused out yielding minimum AOM (Egger et al., 2016). Nevertheless, AOM-SR was observed in the sediment in the presence of anaerobic methane oxidizing communities (Bhattarai et al., 2017). Based on these findings, high rate of SR with commonly used electron donors, such as acetate and ethanol, can be expected, while there could be possible involvement of other sulfur compounds for AOM, besides SO_4^{2-}, e.g. S^0 (Milucka et al., 2012). Therefore, the main objective of this study was to determine the sulfate reducing activities with different electron donors, i.e. ethanol, acetate and lactate in order to compare which one was preferred by the sulfate reducing communities inhabiting the sediment investigated. Further, potential involvement of alternative sulfur compounds (S^0 and $S_2O_3^{2-}$) as electron acceptors for AOM-SR activities were investigated and compared with the AOM-SR rate achieved by SO_4^{2-} as an electron acceptor.

3.2 Materials and Methods

3.2.1 Study site

Marine Lake Grevelingen (MLG) is a former estuary which partly interacts with seawater from the North Sea by dams (Hagens et al., 2015). It receives a high input of organic matter from the North Sea during spring and summer periods. High rates of deposition and degradation of organic matter have resulted in CH_4 rich anoxic sediments, which, when combined with SO_4^{2-} from seawater renders the site a potential niche for SR, including AOM coupled to SR (AOM-SR) (Bhattarai et al., 2016). The lake inhabits unique microbiota, including *Beggiatoa* mats and a novel type of *Desulfobulbus* clade "cable bacteria", in its sediment due to its seasonal hypoxia in the shallow depth and anaerobic organic rich sediment in the deeper part of the lake (Hagens et al., 2015; Sulu-Gambari et al., 2016).

3.2.2 Sampling

Sediment was obtained from the MLG at a water depth of 45 m from the Scharendijke Basin (51° 44.541' N; 3° 50.969' E). The sampling site has the following characteristics: salinity 31.7 ‰, sulfate 25 mM at the surface of the sediment which reduced up to 5 mM at deeper sediment depths (35 cm), sedimentation rate ~ 3 cm yr^{-1} and average temperature 11°C (Egger et al., 2016). The sediment was anaerobic, dark colored with prominent sulfidic odor. On the vessel R/V Luctor in November 2013, coring was done by the Royal Netherlands Institute for Sea Research (Yerseke, the Netherlands). A gravity corer (UWITEC, Mondsee, Austria) was used to collect the sediments, having a core liner internal diameter of 6 cm and a length of 60 cm. The sediment core was sliced every 5 cm and the sediment layer of 10-20 cm depth (dark colored sulfidic sediment) was used for the activity tests.

Sediment was obtained from the MLG at a water depth of 45 m from the Scharendijke Basin (51° 44.541' N; 3° 50.969' E). The sampling site has the following characteristics: salinity 31.7 ‰, O_2 saturation 6.8% and average temperature 11°C (Egger et al., 2016). On the vessel R/V Luctor in November 2013, coring was done by the Royal Netherlands Institute for Sea Research (Yerseke, the Netherlands). A gravity corer (UWITEC, Mondsee, Austria) was used to collect the sediments, having a core liner internal diameter of 6 cm and a length of 60 cm. The sediment was sliced every 5 cm and the sediment layer of 10-20 cm depth (dark colored sulfidic sediment) was used for the activity tests.

3.2.3 Experimental design

The wet sediment was homogenized separately in a N_2-purged anaerobic chamber from PLAS LABS INC^{TM} and diluted with artificial seawater medium in a ratio of 1:3, and then aliquoted in 250 ml sterile serum bottles with 40% headspace. The artificial seawater medium composed of (per liter of demineralized water): NaCl (26 g), KCl (0.5 g) $MgCl_2·6H_2O$ (5 g), NH_4Cl (0.3 g), $CaCl_2·2H_2O$ (1.4 g), KH_2PO_4 (0.1 g), trace element solution (1 ml), 1 M $NaHCO_3$ (30 ml), vitamin solution (1 ml), thiamin solution (1 ml), vitamin B_{12} solution (1 ml), 0.5 g L^{-1} resazurin solution (1 ml) and 0.5 M Na_2S solution (1 ml) (Zhang et al., 2010). 0.0002 g resazurin l^{-1} (redox indicator for anaerobic conditions) was added and the pH was adjusted to 7.0 with sterile 1 M

Na_2CO_3 or 1 M H_2SO_4 solution. pH was adjusted to 7.0 with sterile 1 M Na_2CO_3 or 1 M H_2SO_4 solution, which was stored under nitrogen atmosphere. The medium was kept anoxic through N2 purging until the incubation with the sediment. The prepared serum bottles were incubated in the dark with gentle shaking at room temperature (~ 20 ± 2°C).

Six different activity tests were prepared of various durations according to Table 3.2 with different electron donors (ethanol, lactate, acetate or methane) and different electron acceptors (SO_4^{2-}, $S_2O_3^{2-}$ or S^0). In the case of the ethanol activity test, an additional 5 mM of ethanol was added after 20 days of experiment to test complete SR and then the experiment was continued for 40 days (Table 3.2). SO_4^{2-} and $S_2O_3^{2-}$ were added in the artificial seawater media as Na_2SO_4 (1.43 g) and $Na_2S_2O_3$ (1.58 g) as anhydrous form, both bought from Fisher Scientific (Sheepsbouwersweg, the Netherlands). S^0 was purchased as precipitated sulfur as powder from Fisher Scientific (Sheepsbouwersweg, the Netherlands) and homogenized in the artificial seawater medium by continuous stirring.

Table 3.2 *Incubations set-up, concentration of electron donors and electron acceptors, duration of the incubations and sampling intervals for the SR and AOM activity tests.*

Type of experiment	Electron donor added		Electron acceptor added		Duration (days)	Sampling (days)
	Compound	mM	Compound	mM		
SR activity test[a]	Ethanol[b]	5	SO_4^{2-}	10	40	3
	Lactate	5	SO_4^{2-}	10	30	3
	Acetate	5	SO_4^{2-}	10	30	3
AOM activity test	CH_4	2 bar	SO_4^{2-}	10	225	15
	CH_4 [c]	2 bar	$S_2O_3^{2-}$	10	350	15
	CH_4 [d]	2 bar	S^0	10	350	15

Notes:

[a] Headspace filled with N_2 gas.

[b] Additional 5 mM of ethanol was added after 20 days of experiment to test complete SR and then the experiment was continued for 40 days.

[c] After 250 days, the artificial seawater medium was refreshed with new medium containing 10 mM of $S_2O_3^{2-}$.

[d] After 250 days the artificial sea water medium was refreshed with new medium containing 25 mM of S^0

The samples for microbial analysis by fluorescence in situ hybridization (FISH) and volatile suspended solids (VSS) analysis were collected at the beginning and end of the experiments. The biotic and abiotic controls were prepared in duplicates for each set of experiment. Several control tests were prepared: without electron acceptors, without electron donors, without biomass and with autoclaved (killed) biomass. 2 ml of wet sediment was withdrawn from each bottle, once every three days, for SO_4^{2-} and total sulfide (TS) for all cumulative dissolved sulfide species (H_2S, HS^- and S^{2-}) analysis for the SR activity test with other electron donor than CH_4, while the same amount of slurry was also obtained in an interval of 15 days from the batch incubations for AOM activity test (Table 3.2). The SR activity tests were performed in triplicates, while the AOM activity tests were performed in quadruplets (Table 3.2), then the average and standard deviation of the replicates were calculated.

3.2.4 Rate calculations

The volumetric SR and TS production rates were calculated as described in Eq. 3.1 and 3.4 (Meulepas et al., 2009a):

$$Volumetric\ sulfate\ reduction\ rate = \frac{[SO_4^{2-}{}_{(t)}]-[SO_4^{2-}{}_{(t+\Delta t)}]}{\Delta t} \qquad \text{Eq. 3.1}$$

$$Volumetric\ sulfide\ production\ rate = \frac{[TS_{(t)}]-[TS_{(t+\Delta t)}]}{\Delta t} \qquad \text{Eq. 3.2}$$

Where, $SO_4^{2-}{}_{(t)}$ is the concentration of SO_4^{2-} at time (t) during the batch incubation, $SO_4^{2-}{}_{(t+\Delta t)}$ is the concentration of SO_4^{2-} at time (t+Δt). Similarly, $TS_{(t)}$ is the concentration of TS at time (t) and $TS_{(t+\Delta t)}$ is the TS concentration at time (t+Δt). SO_4^{2-} / TS concentration of maximum gradient in the slope of activity test was considered for the maximum volumetric rate calculation.

$$Specific\ sulfate\ reduction\ rate = \frac{Volumetric\ sulfate\ reduction\ rate}{VSS\ (g)} \qquad \text{Eq. 3.3}$$

$$Specific\ sulfide\ production\ rate = \frac{Volumetric\ sulfide\ production\ rate}{VSS\ (g)} \qquad \text{Eq.}$$

3.4

Where, VSS is the total amount of initial VSS measured in the incubated sediment from MLG, i.e. 16.9 g.

3.2.5 Microbial community analysis

Microbial analysis of biomass was performed by Fluorescence *in situ* hybridization (FISH) method. Sub-samples after 20 days of incubation from batch incubations with ethanol, lactate and acetate were fixed for 2 hours on ice with 3% formaldehyde, centrifuged and washed twice with a phosphate buffer saline (PBS) and stored at -20°C in PBS/ethanol (1:1, *v/v*) prior to FISH analysis. The microorganisms were visualized using archaeal and bacterial specific oligonucleotide probes ARCH915 (Stahl and Amann 1991) and EUB I-III (Daims et al., 1999), respectively, to check the occurrence of archaeal and bacterial cells. *Desulfosarcina/Desulfococcus* (*DSS*) and *Desulfobulbus* (*DBB*), the SO_4^{2-} reducing *Deltaroteobacteria* were visualized by using DSS658 (Boetius et al., 2000) and DBB660 (Daly et al., 2000) probes, respectively. The microbial cells were counterstained with 4', 6-diamidino-2-phenylindole (DAPI) to visualize all microbial cells.

3.2.6 Chemical analysis

The VSS was estimated on the basis of the difference between the dry weight total suspended solids (TSS) and the ash weight of the sediment according to the procedure outlined in Standard Methods (APHA 1995). Dissolved TS was analyzed using the methylene blue method immediately after sampling (Siegel 1965). One volume of sample (0.5 ml) was diluted to one volume of 1 M NaOH to raise the pH in order to prevent the volatilization of sulfide. SO_4^{2-} was analyzed using an Ion Chromatograph system (Dionex-ICS-1000 with AS-DV sampler), as described previously (Villa-Gomez et al., 2011). The pH was measured using a pH indicator paper.

3.3 Results

3.3.1 Sulfate reduction (SR) with ethanol, lactate and acetate as electron donor

The pH at the beginning of the experiments was ~ 7.5 which increased up to 8.8 towards the end of the experiments in the incubations with ethanol, lactate and acetate as electron donors. A similar trend of SR in SO_4^{2-} concentration profiles was observed for the incubations with acetate and ethanol, whereas with lactate the reduction of the SO_4^{2-} occurred within the first 13 days of incubation, after which the SO_4^{2-} concentration remained nearly constant (Figure 3.1).Concomitant with the SO_4^{2-} reduction, all incubations showed an increasing trend of dissolved TS production at the beginning and stable trend towards the end of the incubation period (Figure 3.1).

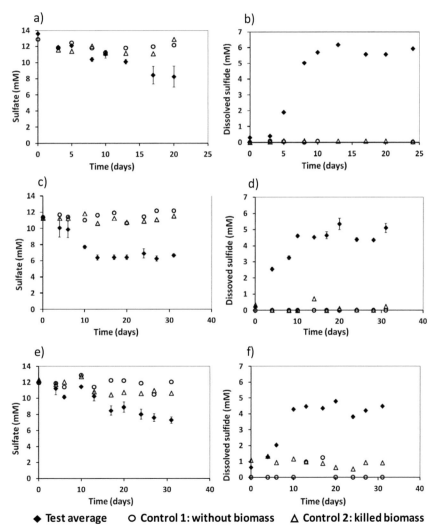

Figure 3.1 Microbial SR activity by Marine Lake Grevelingen (MLG) sediment: a) sulfate consumption with ethanol, b) total sulfide production with ethanol, c) sulfate consumption with lactate, d) total sulfide production with lactate, e) sulfate consumption with acetate and f) total sulfide production with acetate.

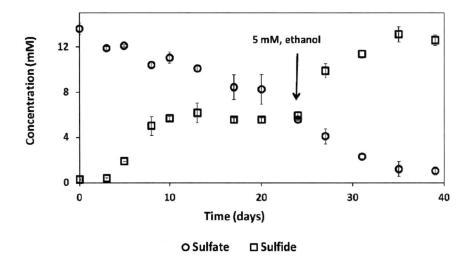

O Sulfate **□** Sulfide

Figure 3.2 *Microbial sulfate reduction (SR) activity by Marine Lake Grevelingen (MLG) sediment with ethanol showing the complete reduction of SO_4^{2-} by the addition of 10 mM of total ethanol in two phases. In the starting 5 mM ethanol and 10 mM of SO_4^{2-} was added to the artificial seawater medium and the incubation was spiked again with 5 mM ethanol after 20 days of incubation.*

Table 3.3 *Rates of sulfate reduction (SR) and total sulfide (TS) production for (Marine Lake Grevelingen) MLG sediment incubations by using different electron donors and different electron acceptors.*

Incubation type	SR rate		TS production rate		SO_4^{2-} removed (mM)	$\%SO_4^{2-}$ removed [c]
	Volumetric (μmol $L^{-1}d^{-1}$)	Specific (μmol $g_{VSS}^{-1}d^{-1}$)	Volumetric (μmol $L^{-1}d^{-1}$)	Specific (μmol $g_{VSS}^{-1}d^{-1}$)		
Ethanol + SO_4^{2-}	920	55	1320	78	6.2 [a]	90
Lactate+ SO_4^{2-}	540	32	580	34.5	5[a]	88
Acetate+ SO_4^{2-}	440	26	560	33	5[a]	78
CH_4 + SO_4^{2-}	80	5	50	3	5.8[b]	50
CH_4 + $S_2O_3^{2-}$	120	7.3	110	7	3.3[b]	33
CH_4 + S^0	18	1	130	7.4	1.6[b]	12

Notes: [a] the result obtained within 30 days of incubation. [b] the result obtained within 160 days of incubation. The rate was estimated on the basis of maximum gradient in the activity slope. [c] % sulfate removed was calculated on the basis of 100% mineralization of the added electron donor.

Among the electron donors studied, the highest SR and TS production rates were observed in the incubation with ethanol, 55 and 78 μmol g_{VSS}^{-1} d^{-1}, respectively (Table 3.3 and Figures 3.1a and 3.1b). In order to test the microbial activity for complete SR and sulfide toxicity, another 5 mM of ethanol was added at around 25 days of incubation to the ethanol batch bottles, after which complete SR of 12 mM of SO_4^{2-} was observed (Figure 3.2).The progress of the experiment showed that actually ethanol was a limiting factor at that point and SR and TS production was increased again by the addition of ethanol.

FISH was performed to observe the microbial cells specifically targeting Archaea, Bacteria, ANME and SRB, which are commonly associated for AOM, i.e., *DSS* and *Desulfobulbus DBB*, after 20 days of incubation with ethanol, lactate or acetate. Diverse cell aggregates of bacteria were observed for all the incubations with acetate, lactate and ethanol (Figure 3.3a). In contrast, archaeal cells could not be visualized for any of the incubation. Clusters of cocci shaped cells of *DBB* were observed with the ethanol incubations only (Figures 3.3b and 3.3c), while no *DSS* were observed.

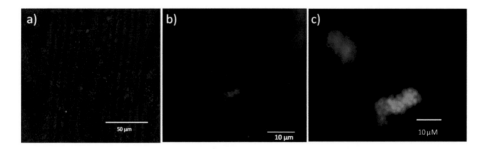

Figure 3.3 Visualization of cells from the activity test with MLG sediment after 20 days of incubation by fluorescence in situ hybridization (FISH). a) All cells stained by 4', 6-diamidino-2-phenylindole (DAPI) in blue and bacterial cells in green hybridized by EUB I-III probe for bacteria, b) all cells stained from the ethanol incubation by DAPI and Desulfobulbus (DBB) cells in green color hybridized by DBB probe and c) green cells showed a group of cocci shaped DBB cells in ethanol incubations.

3.3.2 SR with CH4 as the sole electron donor

In the batch incubations with CH_4 as the sole electron donor with different sulfur compounds, the starting pH was 7.5 and increased up to 8.5 towards the end of the activity test. Dissolved TS in the incubation with CH_4 and SO_4^{2-} was around 6 mM at the end of the experiment, whilst almost 7.5 mM SO_4^{2-} was consumed (Figure 3.4a). The SR rate for the incubation with CH_4 was much higher compared to the SR rate obtained in control incubations, i.e. without methane and without biomass (Figure 3.5).

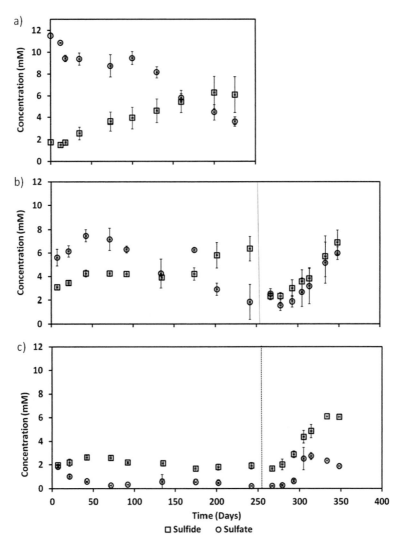

Figure 3.4 *Microbial SR activity by MLG sediment with CH_4 as electron donor and a) 10 mM of SO_4^{2-} (the error bar indicates standard deviation, n=4), b) 10 mM $S_2O_3^{2-}$ (the error bar indicates standard deviation, n=4) and c) 10 mM S^0 as electron acceptor (the error bar indicates standard deviation, n=3). Dashed line indicates the replacement of the medium with fresh artificial seawater medium, headspace was flushed with CH_4 and addition of b) 10 mM $S_2O_3^{2-}$ and c) 25 mM S^0.*

Similarly, the TS concentration for the incubations without the biomass and with CH_4 and SO_4^{2-} was almost zero during the incubation period and for the incubation without CH_4 it was almost three times less than the cumulative TS concentration for the incubation with CH_4 and SO_4^{2-}. Trace organic matter utilization by the SRB might have contributed to the dissolved TS production during the initial periods (100 days) of incubation.

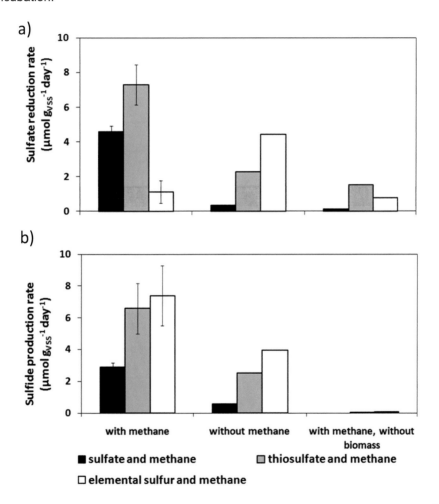

Figure 3.5 *Maximum specific a) sulfate reduction and b) total sulfide production rates (μmol g_{VSS}^{-1} d^{-1}) for AOM activity test of MLG sediment with different electron acceptors (SO_4^{2-}, $S_2O_3^{2-}$ and S^0) and CH_4 as electron donor and control incubations without CH_4 and without biomass. The error bars indicate standard deviation (n=4 for SO_4^{2-} and $S_2O_3^{2-}$) and (n=3 for S^0). Here, the initial VSS) of MLG sediment (16.9 g L^{-1}) was used for the calculation of the specific rate for the samples with biomass. For the abiotic control, the rate was estimated considering no biomass.*

In the incubations with $S_2O_3^{2-}$ (Figure 3.4b) both SO_4^{2-} and dissolved TS concentrations reached up to 7.8 mM and 4.2 mM, respectively during the first 50 days. After 150 days of incubations, dissolved TS increased to 6.4 mM, while SO_4^{2-} was reduced from 7.8 to 1.8 mM (Figure 3.4b). After day 250, the batches were refreshed by 10 mM $S_2O_3^{2-}$ containing saline mineral medium and pressurized with 2 bar of CH_4. Then, both dissolved TS and SO_4^{2-} increased exponentially to 7 mM and 6 mM, respectively, until the end of the experiment. Dissolved TS production and SO_4^{2-} consumption was not observed in control incubations in abiotic incubations. The results from control incubation without CH_4 with $S_2O_3^{2-}$ showed that the SR and dissolved TS production rates were 3 times lower than those observed in the incubation with CH_4 (Figure 3.5). However, the SR rate (2.3 µmol g_{VSS}^{-1} day^{-1}) for the control without CH_4 with $S_2O_3^{2-}$ was much higher than the SR rate (0.1 µmol g_{VSS}^{-1} day^{-1}) for control incubation with SO_4^{2-} and the absence of CH_4 (Figure 3.5).

In the incubations with S^0, consumption of SO_4^{2-} and dissolved TS production was observed only during the first 50 days, however less in amount compared to the incubations with $SO_4^{2-}/S_2O_3^{2-}$ (Figure 3.4c). Upon replacement of the mineral medium after 250 days of incubation, both SO_4^{2-} and dissolved TS levels increased abruptly and reached 2.7 mM and 4.2 mM, respectively, at around day 320. After 350 days of incubation, SO_4^{2-} was almost completely consumed and the dissolved TS levels increased to 6 mM. TS production and SO_4^{2-} consumption was not observed in the abiotic control incubations with S^0. The control incubation without CH_4 showed a similar SR rate as in the incubation with CH_4 and S^0 (Figure 3.5).

In this study with $S_2O_3^{2-}$ and S^0 incubations, methane consumption was not observed; nevertheless, CO_2 production was almost similar for the activity test incubations with CH_4 and control incubation without CH_4. Therefore, net AOM could not be estimated when $S_2O_3^{2-}$ or S^0 were used as electron acceptor.

3.4 Discussion

3.4.1 Sulfate reduction (SR) by Marine Lake Grevelingen (MLG) sediment with different electron donors

SR by the microbiota present in MLG sediment was faster with ethanol as the substrate compared to the other electron donors tested. The SR rates with different electron donors obtained in this study were almost 100-200 times lower than those obtained by anaerobic granular sludge originating from bioreactors (Hao et al., 2014; Liamleam and Annachhatre 2007). However, VSS from the sediment might overestimate the microbial biomass in the sediment, since it can include both cell biomass and organic matter present in the sediment. Therefore, SR rates determined in this study might be lower than those obtained by anaerobic granular sludge originating from bioreactors. The SR rates with ethanol, lactate and acetate (Table 3.3) were, nevertheless, higher compared to the SR rates in marine coastal sediments from other shallow coastal sediments, such as Eckernförde Bay sediment with a water depth of 20 m, ranging between 0.020 and 0.465 mmol L^{-1} d^{-1} (Treude et al., 2005) or organic-rich shallow sediment of Limfjorden with a water depth of 10 m (eutrophic sound in Denmark connecting to the North Sea), ranging between 0.001 and 0.1 mmol L^{-1} d^{-1} (Jorgensen and Parkes 2010). Nevertheless, these measurements were performed in short term incubations for around five days at the *in situ* temperature ranging from 9°C to 13°C. In this study the rate was measured over a period of 20 days (for ethanol, lactate and acetate) and more than 200 days (for CH_4) at 20 (±2)°C.

All incubations with different electron donors (ethanol, lactate and acetate) and SO_4^{2-} showed simultaneous dissolved TS production and SR. Consumption of ethanol, acetate and lactate were almost instantaneous in the different incubations and the maximum dissolved TS concentration was obtained within ~ 10 days of incubation (Figure 3.1). Thereafter, the maximum dissolved TS concentration remained stable which was due to the lack of electron donor as the SR activity resumed after another ethanol addition (Figure 3.2).

The microbial community in the MLG sediment was active at dissolved TS concentrations exceeding 10 mM. SRB have a wide range of TS tolerance, up to ~

16 mM of dissolved TS present in the incubation medium (Reis et al., 1992). A detailed study of dissolved TS toxicity onto marine AOM-SR consortia is still lacking. Nevertheless, with sediment from the Gulf of Mexico in the active seepage area, a dissolved TS concentration up to 12 mM was observed (Joye et al., 2004) and accumulation of 14 mM dissolved TS was observed in the incubation of sediment hosting AOM from hydrate ridge (Nauhaus et al., 2005). In contrast, dissolved TS toxicity was observed already at 2 mM of dissolved TS with coastal estuarine sediment from Eckernförde bay (Meulepas et al., 2009a).

All three electron donors used in this study are utilized by a wide range of SRB. Lactate and ethanol can be fermented to short chain volatile fatty acids (VFA), such as acetate, in the presence of SO_4^{2-} and then only oxidized to bicarbonate (Zellner et al., 1994). Typically, ethanol and lactate are metabolized by *Desulfovibrio, Desulfomonas, Desulfotomaculum DBB* and *Desulfotomaculum* species of SRB to VFA or hydrogen (Muyzer and Stams 2008; Bryant et al., 1977). Acetate is mainly utilized by *Desulfobacter, Desulfococcus, DSS,* and *Desulfonema* clades of the SRB (Martinko and Madigan 2005). Thermodynamically, SR coupled to acetate oxidation releases the highest energy, whereas SR in the presence of ethanol or lactate has less negative Gibb's free energy values (Table 3.1). Further, acetate is considered as major substrate for SR in marine and estuarine sediments (Parkes et al., 1989). However, the result from this study showed the lowest SR rate with acetate (Table 3.3), which suggests that acetate could have been used for other processes. For instance, if methanogens are present in the MLG sediment, competition can occur between SRB and methanogens for acetate utilization, leading to lower SR. A detailed analysis of the microbial communities along with SRB diversity could be performed in future studies to link the microbiome with their carbon source and electron donor utilization as well as their metabolic pathways.

3.4.2 AOM with different sulfur compounds

MLG sediment is able to utilize all three electron acceptors, i.e., SO_4^{2-}, $S_2O_3^{2-}$ and S^0, with CH_4 as the electron donor (Figure 3.4). While assessing the SR with $S_2O_3^{2-}/S^0$, the active sulfur disproportionated TS production was observed with these sulfur compounds instead of AOM induced TS production. CO_2 measurements for the incubations with S^0 and $S_2O_3^{2-}$ did not clearly show the oxidation of CH_4 to CO_2, as

CH_4 remained constant and CO_2 was produced in both incubations with and without CH_4 in the headspace. Nevertheless, it was observed that the CH_4 consumption was ~ 5.5 mM with the simultaneous production of 1.5 mM CO_2 in the batch incubations with CH_4 and SO_4^{2-} (Bhattarai et al., 2017).

Similar to this study, SR rate (50 to 80 µmol SO_4^{2-} l^{-1} day^{-1}) was observed with Eckernförde Bay sediment in the beginning of an enrichment experiment in a bioreactor (Meulepas et al., 2009a). The observed dissolved TS production rate in this study with all electron acceptor were higher compared to the rate obtained after incubation of the cold seep sediment from Captain Aryutinov Mud Volcano, Gulf of Cadiz at 2 bar pressure (0.18 µmol TS g_{dw}^{-1} day^{-1}) using CH_4 as electron donor and SO_4^{2-} as an electron acceptor (Zhang et al., 2010). Similarly, a mixture of coastal sediments from the Aarhus bay and Eckernförde bay using CH_4 and other alkanes as electron donors with SO_4^{2-} (1.5 to 2.5 µmol TS l^{-1} day^{-1}) had lower dissolved TS production rates than those observed in this study (Suarez-Zuluaga et al., 2014).

The SR rate obtained in the incubation with CH_4 and $S_2O_3^{2-}$ was comparatively higher to the SR rate obtained from the parallel incubation with CH_4 and SO_4^{2-}. Moreover, the dissolved TS production rates with CH_4 and $S_2O_3^{2-}$ or S^0 were higher (110 and 130 µmol TS l^{-1} day^{-1} respectively) compared to the TS production rates (80 µmol TS l^{-1} day^{-1} for $S_2O_3^{2-}$ and 1.2 µmol TS l^{-1} day^{-1} for S^0) of the mixture of Aarhus bay and Eckernförde bay sediment (Suarez-Zuluaga et al., 2014). The SR process with $S_2O_3^{2-}$ was indirectly activated by disproportionation rather than AOM induced SR. Previous studies indicated that the alkane degradation by marine sediments might be facilitated by the enrichment of alkane degraders with the addition of $S_2O_3^{2-}$ (Meulepas et al., 2009b ; Suarez-Zuluaga et al., 2014). The comparatively higher rate of SR in the control incubation without methane for the case of $S_2O_3^{2-}$ or S^0 might be caused by the rapid enrichment of sulfate reducing bacteria and degradation of residual organic matter. Benthic organic matter contains various fractions among which a small portion can be degraded quickly and a major portion is more recalcitrant (Arndt et al., 2013). The latter might, nevertheless, have been degraded during the long term incubation of the control without methane fueling SR in these incubations.

The highest SR rate was observed in the batch tests with CH_4 and $S_2O_3^{2-}$ after the first 50 days of incubations (7.3 SO_4^{2-} μmol g_{VSS}^{-1} d^{-1}), suggesting that SO_4^{2-} was produced due to disproportionation and then reduced to sulfide concomitant to CH_4 oxidation and organic matter degradation (Figure 3.4). Theoretically more energy can be released using different oxidized forms of sulfur than SO_4^{2-}, such as $S_2O_3^{2-}$ (Table 3.1), which might lead to higher AOM rates (Suarez-Zuluaga et al., 2014). The fast SO_4^{2-} production by $S_2O_3^{2-}$ might trigger high SRB activity and consequent high SR rate.

However, the disproportionation of both S^0 and $S_2O_3^{2-}$ to SO_4^{2-} and sulfide was predominant (Figure 3.2). S^0 and $S_2O_3^{2-}$ are important intermediates during sulfide oxidation in marine sediments (Fossing and Jorgensen 1990). The disproportionation of $S_2O_3^{2-}$ is energetically favorable and the disproportionation of S^0 requires energy unless an oxidant, as Fe (III), renders the reaction more energetically favorable (Finster 2008) or in alkaline environments, such as soda lakes (Poser et al., 2013). Similarly, high rates of $S_2O_3^{2-}$ disproportionation were reported in a study with coastal marine sediment and CH_4 as the sole carbon source (Suarez-Zuluaga et al., 2015). In that study, a high number of *Desulfocapsa* was observed, which are specialized in disproportionation of sulfur compounds (Finster et al., 1998). Further, pH of the sediment from MLG ranged between 7.6 and 8.4 (Hagens et al., 2015), while the amount of Fe oxides in the sediment ranged between 20 and 50 μmol g^{-1} (Egger et al., 2016). Therefore, microbial S^0 disproportionation might have been possible due to Fe (III) acting as sulfide scavenger, e.g. by bacteria such as *Desulfocapsa* (Finster 2008) as majority of these disproportionating bacteria need Fe (III) or Mn (IV) as sulfide scavenger. Alternativley, S^0 disproportionation could have been via the metabolism of haloalkaliphilic bacteria, which can disproportionate S^0 without Fe(III) or Mn(IV). However, these bacteria are commonly found in soda lakes (Poser et al., 2013) and likely do not occur in the MLG sediment. In order to differentiate and explain among these mechanisms, the archaeal and bacterial community inhabiting the MLG sediment should be further studied by e.g. genome analysis and Fluorescence *In Situ* Hybridization (FISH) techniques as Catalyzed reporter deposition FISH (CARD-FISH) and FISH with micro radiography (FISH-MAR).

CH_4 could be utilized by the microbial community present in the MLG sediment (Bhattarai et al., 2017). The MLG sediment is capable to perform SR by using diverse electron donors, including the gaseous and cheap electron donor methane. S^0 and $S_2O_3^{2-}$ as electron acceptors showed sulfur disproportionation possibly by SRB, however, $S_2O_3^{2-}$ could be used as a trigger for faster SR. Further, the SO_4^{2-} reducing microbial community in the studied sediment is active at high TS concentrations (Figure 3.2) and they are comparable to previously studied methane seep sediments (Zhang et al., 2010). Therefore, the MLG sediment can be used for further enrichment in bioreactors. Moreover, investigations of the SRB and ANME species responsible for AOM-SR and of sulfur disproportionation are required to exploit their potential application in the field of environmental biotechnology.

3.5 Conclusions

This study aimed to explore microbial sulfate reducing activity of coastal sediment from the MLG using different electron donors and electron acceptors, including gaseous electron donor i.e., methane. The SO_4^{2-} reducing microbial community in the sediment was capable to utilize all four different tested electron donors namely, ethanol, lactate, acetate and methane. The SR activity with methane indicated the microbial capability for AOM. Moreover, when $S_2O_3^{2-}$ and S^0 were supplied along with methane to incubations instead of SO_4^{2-}, SR activity was observed in all the cases, mainly by disproportionation of sulfur compound, rather than AOM. Among the three tested sulfur compounds used as an electron acceptor with methane, the higher SR rate was observed with $S_2O_3^{2-}$ · though via disproportionation. Therefore, the use $S_2O_3^{2-}$ in the initial phase of bioreactor operation may activate the faster rate of SR. Thus, this study widens our understanding on potential use of marine sediments with diverse microbial clade in SO_4^{2-} removal from wastewater.

3.6 References

APHA. 1995. Standard methods for the examination of water and wastewater, *American Public Health Association*. (19th edition), Washington DC, USA: p 1325.

Arndt S, Jørgensen B.B., LaRowe D.E., Middelburg J, Pancost R, Regnier P. 2013. Quantifying the degradation of organic matter in marine sediments: a review and synthesis. *Earth-Sci. Rev.***123**,53-86.

Bhattarai, S., Cassarini, C., Gonzalez-Gil, G., Egger, M., Slomp, C.P., Zhang, Y., Esposito, G., Lens, P.N.L. 2017. Anaerobic methane-oxidizing microbial community in a coastal marine sediment: anaerobic methanotrophy dominated by ANME-3. *Microb. Ecol.*, **74**(3), 608-622.

Boetius, A., Ravenschlag, K., Schubert, C.J., Rickert, D., Widdel, F., Gieseke, A., Amann, R., Jorgensen, B.B., Witte, U., Pfannkuche, O. 2000. A marine microbial consortium apparently mediating anaerobic oxidation of methane. *Nature*, **407**(6804), 623-626.

Bryant M.P., Campbell L.L., Reddy C.A., Crabill M.R. 1977. Growth of *Desulfovibrio* in lactate or ethanol media low in sulfate in association with H_2-utilizing methanogenic bacteria. *Appl. Environ. Microbiol.* **33**(5), 1162-1169.

Daims H, Brühl A, Amann R, Schleifer K-H, Wagner M. 1999. The domain-specific probe EUB338 is insufficient for the detection of all Bacteria: development and evaluation of a more comprehensive probe set. *Syst. Appl. Microbiol.* **22**(3), 434-444.

Daly K, Sharp R.J., McCarthy A.J. 2000. Development of oligonucleotide probes and PCR primers for detecting phylogenetic subgroups of sulfate-reducing bacteria. *Microbiology* **146**(7),1693-1705.

Deusner, C., Meyer, V., Ferdelman, T. 2009. High-pressure systems for gas-phase free continuous incubation of enriched marine microbial communities performing anaerobic oxidation of methane. *Biotechnol. Bioeng.*, **105**(3), 524-533.

D'Hondt, S., Jørgensen, B.B., Miller, D.J., Batzke, A., Blake, R., Cragg, B.A., Cypionka, H., Dickens, G.R., Ferdelman, T., Hinrichs, K.-U. 2004. Distributions of microbial activities in deep subseafloor sediments. *Science*, **306**: 2216-2221.

Finster K. 2008. Microbiological disproportionation of inorganic sulfur compounds. *J. Sulfur. Chem.* **29**(3-4), 281-292.

Finster K, Liesack W, Thamdrup B. 1998. Elemental sulfur and thiosulfate disproportionation by *Desulfocapsa sulfoexigens* sp . nov., a new anaerobic bacterium isolated from marine surface sediment. *Appl. Environ. Microbiol.* **64**(1), 119-125.

Fossing H, Jorgensen B.B. 1990. Oxidation and reduction of radiolabeled inorganic sulfur compounds in an estuarine sediment, Kysing Fjord, Denmark. *Geochim. Cosmochim. Ac.* **54**(10), 2731-2742.

Gonzalez-Gil, G., Meulepas, R.J.W., Lens, P.N.L. 2011. Biotechnological aspects of the use of methane as electron donor for sulfate reduction. in: Murray, M.-Y. (Ed.),*Comprehensive Biotechnology*. Vol. 6. (2nd edition), Elsevier B.V., Amsterdam, the Netherlands pp. 419-434.

Hagens M, Slomp C.P., Meysman F.J.R., Seitaj D, Harlay J, Borges A.V., Middelburg J.J. 2015. Biogeochemical processes and buffering capacity concurrently affect acidification in a seasonally hypoxic coastal marine basin. *Biogeosciences* **12**(5), 1561-1583.

Hao T-w, Xiang P-y, Mackey HR, Chi K, Lu H, Chui H-k, van Loosdrecht M.C.M., Chen G-H. 2014. A review of biological sulfate conversions in wastewater treatment. *Water Res.* **65**, 1-21.

Heimann A.C., Friis A.K., Jakobsen R. 2005. Effects of sulfate on anaerobic chloroethane degradation by an enriched culture under transient and steady-state hydrogen supply. *Water Res.* **39**, 3579-3586.

Jorgensen B.B., Parkes R.J. 2010. Role of sulfate reduction and methane production by organic carbon degradation in eutrophic fjord sediments (Limfjorden, Denmark). *Limnol. Oceanogr.* 53(3),1338-1352.

Joye, S.B., Boetius, A., Orcutt, B.N., Montoya, J.P., Schulz, H.N., Erickson, M.J., Lugo, S.K. 2004. The anaerobic oxidation of methane and sulfate reduction in sediments from Gulf of Mexico cold seeps. *Chem. Geol.*, **205**(3), 219-238.

Knittel, K., Boetius, A. 2009. Anaerobic oxidation of methane: progress with an unknown process. *Annu. Rev. Microbiol.*, **63**(1), 311-334.

Krüger, M., Blumenberg, M., Kasten, S., Wieland, A., Känel, L., Klock, J.-H., Michaelis, W., Seifert, R. 2008. A novel, multi-layered methanotrophic microbial mat system growing on the sediment of the Black Sea. *Environ. Microbiol.*, **10**(8), 1934-1947.

Liamleam W, Annachhatre A.P. 2007. Electron donors for biological sulfate reduction. *Biotechnol. Adv.*, **25**(5), 452-463.

Martinko, J. M., & Madigan, M. T. 2005. Brock biology of microorganisms. in: Englewood Cliffs (Ed.), 11th edition, NJ Prentice Hall, New Jersy, USA.

Meulepas, R.J.W., Stams, A.J.M., Lens, P.N.L. 2010. Biotechnological aspects of sulfate reduction with methane as electron donor. *Rev. Environ. Sci. Biotechnol.*, **9**(1), 59-78.

Meulepas, R.J.W., Jagersma, C.G., Gieteling, J., Buisman, C.J.N., Stams, A.J.M., Lens, P.N.L. 2009a. Enrichment of anaerobic methanotrophs in sulfate-reducing membrane bioreactors. *Biotechnol. Bioeng.*, **104**(3), 458-470.

Meulepas, R.J.W., Jagersma, C.G., Khadem, A.F., Buisman, C.J.N., Stams, A.J.M., Lens, P.N.L. 2009b. Effect of environmental conditions on sulfate reduction with methane as electron donor by an Eckernförde Bay enrichment. *Environ. Sci. Technol.*, **43**(17), 6553-6559.

Muyzer, G., Stams, A.J. 2008. The ecology and biotechnology of sulphate-reducing bacteria. *Nat. Rev. Microbiol.*, **6**(6), 441-454.

Nauhaus, K., Albrecht, M., Elvert, M., Boetius, A., Widdel, F. 2007. *In vitro* cell growth of marine archaeal-bacterial consortia during anaerobic oxidation of methane with sulfate. *Environ. Microbiol.*, **9**(1), 187-196.

Nauhaus, K., Treude, T., Boetius, A., Kruger, M. 2005. Environmental regulation of the anaerobic oxidation of methane: a comparison of ANME-I and ANME-II communities. *Environ. Microbiol.*, 7(1), 98-106.

Parkes R.J., Gibson G., Mueller-Harvey I., Buckingham W., Herbert R. 1989. Determination of the substrates for sulphate-reducing bacteria within marine and esturaine sediments with different rates of sulphate reduction. *Microbiology*, **135**(1), 175-187.

Poser A., Lohmayer R., Vogt C., Knoeller K., Planer-Friedrich B., Sorokin D., Richnow H-H, Finster K. 2013. Disproportionation of elemental sulfur by haloalkiphilic bacteria from soda lakes. *Extremophiles*, **17**(6), 1003-1012.

Reis M.A.M, Almeida J.S., Lemos P.C., Carrondo M.J.T. 1992. Effect of hydrogen sulfide on growth of sulfate reducing bacteria. *Biotechnol. Bioeng.*, **40**(5), 549-642.

Siegel L.M. 1965. A direct microdetermination for sulfide. *Anal. Biochem.* **11**(1), 126-132.

Stahl DA, Amann RI. 1991. Development and application of nucleic acid probes. in: Stackebrandt, E., Goodfellow, M. (Eds.), *Nucleic acid techniques in bacterial systematics.* John Wiley & Sons Ltd, Chichester, UK, pp 205-248.

Suarez-Zuluaga D, Timmers P.H.A, Plugge C.M., Stams A.J.M., Buisman C.J.N, Weijma J. 2015. Thiosulphate conversion in a methane and acetate fed membrane bioreactor. *Environ. Sci. Pollut. Res.*, **23**(3), 2467-2478.

Suarez-Zuluaga D, Weijma J, Timmers P.H.A, Buisman C.J.N. 2014. High rates of anaerobic oxidation of methane, ethane and propane coupled to thiosulphate reduction. *Environ. Sci. Pollut. Res.*, **22**(5), 3697-3704.

Sulu-Gambari F, Seitaj D, Meysman F.J.R., Schauer R, Polerecky L, Slomp C.P. 2016. Cable bacteria control iron-phosphorous dynamics in sediments of a coastal hypoxic basin. *Environ. Sci. Technol.*, **50**(3), 1227-1233.

Thauer R.K., Jungermann K, Decker K. 1977. Energy conservation in chemotrophic anaerobic bacteria. *Bacteriol. Rev.*, **41**(1), 100-809.

Treude, T., Krüger, M., Boetius, A., Jørgensen, B.B. 2005. Environmental control on anaerobic oxidation of methane in the gassy sediments of Eckernförde Bay (German Baltic). *Limnol. Oceanogr.*, **50**(6), 1771-1786.

Vasquez-Cardenas, D., van de Vossenberg, J., Polerecky, L., Malkin, S.Y., Schauer, R., Hidalgo-Martinez, S., Confurius, V., Middelburg, J.J., Meysman, F.J.R., Boschker, H.T.S. 2015. Microbial carbon metabolism associated with electrogenic sulphur oxidation in coastal sediments. *ISME J.*, **9**(9), 1966-1978.

Villa-Gomez D, Ababneh H, Papirio S, Rousseau D.P.L., Lens P.N.L. 2011. Effect of sulfide concentration on the location of the metal precipitates in inversed fluidized bed reactors. *J. Hazard. Mater.*, **192**(1), 200-207.

Weijma J, Veeken A, Dijkman H, Huisman J, Lens P. 2006. Heavy metal removal with biogenic sulphide: advancing to full-scale. in: Cervantes, F.J.,

Pavlostathis S.G., van Handeel, A.C. (Eds.) *Advanced biological treatment processes for industrial wastewaters, principles and applications.* Intregated Environmental Technology Series, IWA publishing, London, pp 321-331.

Xiong, Z.-Q., Wang, J.-F., Hao, Y.-Y., Wang, Y. 2013. Recent advances in the discovery and development of marine microbial natural products. *Mar. Drugs.*, **11**:700-717.

Zellner G, Neudörfer F, Diekmann H. 1994. Degradation of lactate by an anaerobic mixed culture in a fluidized-bed reactor. *Water. Res.*, **28**(6), 1337-1340.

Zhang, Y., Henriet, J.-P., Bursens, J., Boon, N. 2010. Stimulation of in vitro anaerobic oxidation of methane rate in a continuous high-pressure bioreactor. *Bioresour. Technol.*, **101**(9), 3132-3138.

CHAPTER 4

Anaerobic Methane Oxidizing Microbial Community in a Coastal Marine Sediment: Anaerobic Methanotrophy Dominated by ANME-3

The modified version of this chapter was published as:

Bhattarai S., Cassarini C., Gonzalez-Gil G., Egger M., Slomp C.P., Zhang Y., Esposito G.and Lens P.N.L. (2017) Anaerobic methane-oxidizing microbial community in a coastal marine sediment: anaerobic methanotrophy dominated by ANME-3. *Microbial Ecology,* 74(3), 608-622

Abstract

The microbial community inhabiting the shallow sulfate-methane-transition-zone in coastal sediments from Marine Lake Grevelingen (the Netherlands) was characterized, and the ability of the microorganisms to carry out anaerobic oxidation of methane coupled to sulfate reduction was assessed in activity tests. *In vitro* activity tests of the sediment with methane and sulfate demonstrated sulfide production coupled to the simultaneous consumption of sulfate and methane at approximately equimolar ratios over a period of 150 days. The maximum sulfate reduction rate was 5 µmol sulfate per gram dry weight per day during the incubation period. Diverse archaeal and bacterial clades were retrieved from sequence analysis with the majority of them clustering to *Euryarchaeota*, *Thaumarcheota*, *Bacteroidetes* and *Proteobacteria*. The 16S rRNA sequence analysis showed that the sediment from Marine Lake Grevelingen contained anaerobic methanotrophic *Archaea* (ANME) and methanogens as archaeal clades with a role in the methane cycling. ANME at the studied site mainly belong to the ANME-3 clade. Sulfate reducing bacteria from *Desulfobulbus* clades were found among the sulfate reducers, however with very low relative abundance. *Desulfobulbus* is commonly associated with ANME, but ANME and *Desulfobulbus* were not visualized together, suggesting the possibility of independent AOM by ANME-3. This study provides one of the few reports for the presence of ANME-3 in a shallow coastal marine sediment.

4.1   Introduction

Methane production and consumption in organic rich marine sediments are largely regulated by microbial processes (Reeburgh, 2007). In oxygenated sediments, methane is scavenged by aerobic oxidation (Tavormina et al., 2010), while anaerobic oxidation of methane (AOM) is responsible for methane consumption in anoxic marine sediments (Knittel & Boetius, 2009; Reeburgh, 2007; Ruff et al., 2016; Wankel et al., 2012). AOM can be coupled different electron acceptors, namely sulfate (Knittel & Boetius, 2009), iron or manganese (Beal et al., 2009), nitrite (Ettwig et al., 2010) and nitrate (Haroon et al., 2013).

AOM coupled to sulfate reduction is carried out by a specific group of anaerobic methanotrophic *Archaea* (ANME) and sulfate reducing bacteria. A very low amount of Gibbs free energy is theoritically released by AOM coupled to sulfate reduction ($\Delta G^{0'}$ of -17 or -34 kJ mol^{-1} with methane as gas or in aqueous form, respectively) to sustain the metabolism of both the ANME and sulfate reducing bacteria (Boetius et al., 2000; Hoehler et al., 1994; Milucka et al., 2012). ANME are grouped into three distinct clades, i.e., ANME-1, ANME-2 and ANME-3 based on the phylogenetic analysis of their 16S rRNA genes (Boetius et al., 2000; Hinrichs et al., 1999; Knittel et al., 2005; Niemann et al., 2006a; Sivan et al., 2007). In the phylogenetic tree, ANME-2 and ANME-3 are clustered with the order of the *Methanosarcinales*, whereas ANME-1 forms a separate cluster that has only a distant relationship to the order *Methanomicrobiales*. The common sulfate reducing bacteria that are associated with ANME belong to the *Desulfosarcina / Desulfococcus* clades and *Desulfobulbaceae*. With a few exceptions, ANME-1 and ANME-2 are commonly found in association with *Desulfosarcina / Desulfococcus* (Boetius et al., 2000), whereas ANME-3 clades are generally observed together with the *Desulfobulbaceae* group of the sulfate reducing bacteria (Niemann et al., 2006b).

AOM coupled to sulfate reduction occurs in a wide range of marine locations, such as cold seeps (Boetius & Wenzhofer, 2013; Orphan et al., 2004; Ruff et al., 2016), mud volcanoes (Niemann et al., 2006a), deep-sea hydrothermal vents (Biddle et al., 2011; Brazelton et al., 2006) and deep-sea sediments (Boetius et al., 2000; Sivan et al., 2007). Environmental conditions such as temperature play a key role in the distribution of ANME. ANME-1 are observed in environments subject to a wide range of temperatures, from 4°C to 70°C (Holler et al., 2011; Vigneron et al., 2013a), whereas ANME-2 and ANME-3 are mostly found in environments with temperatures < 20°C (Hinrichs et al., 2000; Losekann et al., 2007; Niemann et al., 2006b; Orphan et al., 2001). ANME-1 and ANME-2 are the dominant ANME phylotypes in shallow estuarine sediments (Parkes et al., 2007; Treude et al., 2005b). ANME-3 so far appears to be mainly restricted to submarine mud volcanoes and has only been sporadically found in other types of seep sediments (Knittel & Boetius, 2009; Vigneron et al., 2013a). ANME-3 have been found, however, in sediments from the Helgoland mud area (North Sea) at depths in the methanogenic zone where dissolved iron concentrations were high suggesting a potential role of ANME-3 in

AOM and iron reduction (Oni et al., 2015). The reasons for the limited distribution of ANME-3 and details regarding the environmental conditions that determine the prevalence of this type of ANME are still poorly understood and no draft genome of this ANME type is available until now.

Many of ANME habitats that have been studied to date are located in continental slope and deep-sea environments including seeps and mud volcanoes (i.e., at water depths greater than 100 m) (Boetius & Wenzhofer, 2013; Joye et al., 2004; Marlow et al., 2014; Michaelis et al., 2002; Niemann et al., 2006b; Orcutt et al., 2005; Orphan et al., 2004; Treude et al., 2007). Considering the limited accessibility of such ANME habitats (Thornburg et al., 2010; Thurber et al., 2014) and logistic difficulties to cultivate them in the laboratory at *in situ* conditions (Boetius et al., 2009; Meulepas et al., 2010; Zhang et al., 2011), further exploration of ANME in coastal sediments (Dale et al., 2008; Oni et al., 2015; Treude et al., 2005b) is of interest. Shallow coastal systems enable a more easy access to AOM hosting sediments. Further, ANME retrieved from coastal sediments can potentially be enriched under atmospheric pressure (e.g. [Meulepas et al., 2009]). However, although recent ANME studies have been extended to coastal sediments (e.g. [Oni et al., 2015]), exploration of different ANME types remains difficult due to limited primer coverage and methodological biases.

Here, we assess the microbial community composition related to methane cycling of methane-rich coastal sediment of Marine Lake Grevelingen, the Netherlands (max. water depth of 45 m). Marine Lake Grevelingen has a salinity of 32 per mil, is eutrophic (Hagens et al., 2015) and receives a high input of organic matter from the North Sea during spring and summer. This high input has led to the development of a shallow sulfate-methane-transition-zone (SMTZ) in the deeper parts of the lake (Egger et al., 2016), which provides a niche for unique microbiota in the sediment. In the lake sediment, a major proportion of the methane may be bypassing the zone with high sulfate concentrations without being oxidized because of the high rate of sediment accumulation (Egger et al., 2016). Given the similarities of the depositional setting at the study site with a cold seep, the presence of different ANME types, especially ANME-2 and ANME-3, was hypothesized. To assess whether this was indeed the case, the AOM activity with sulfate as an electron acceptor was tested with sediment from the putative SMTZ, whereas the identification and visualization of

110

the ANME type present in the sediment was done by sequencing and fluorescence *in situ* hybridization (FISH).

4.2 Materials and methods

4.2.1 Site description and sampling procedure

Marine Lake Grevelingen is a former estuary, located in the southwestern part of the Netherlands that was transformed into a lake when it was dammed at its eastern and western side in 1964 and 1971 (Figure 4.1). The lake experiences seasonal variations in bottom water temperature between ~ 1.5 to 17°C (Hagens et al., 2015; Sulu-Gambari et al., 2015). Rates of organic matter accumulation, sulfate reduction and methane production in the deep basin sediments are high (Egger et al., 2016). The sediments are dark colored and have a strong smell of sulfide. An unusually high sedimentation rate of around 13 cm yr^{-1} was estimated for this study site based on geochemical profiles (Egger et al., 2016). The water column is hypoxic or anoxic during summer stratification, but fully oxygenated in the winter, which was the time of sampling (Egger et al., 2016; Hagens et al., 2015). The oxygen penetration into the sediment during winter is limited to the upper few millimeters (Meysman, personal communication).

Pore water data from February 2013 and November 2012 as presented by Egger et al. (2016) are included here to provide the geochemical context for this study. For additional geochemical profiles from cores collected between 2012 and 2015, the reader is kindly referred to Egger et al. (2016). Two sediment cores for microbial analysis and AOM activity tests were obtained in December 2013 at the same location (51.742N, 3.849E) as that of Egger et al. (2016). The sediment layers at 5-15 cm depth in each core were sampled under a nitrogen atmosphere, mixed together and a homogenized slurry mix was prepared. A part of this sediment slurry was stored at -20 °C for microbial community analysis and the rest of the sediment slurry was used for the activity tests.

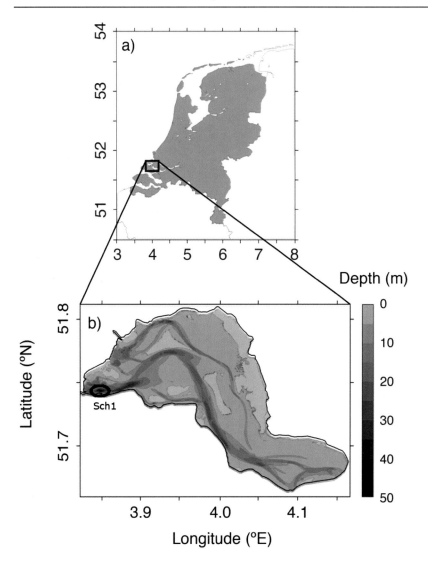

Figure 4.1. *Bathymetry of Marine Lake Grevelingen (the Netherlands) with sampling station 'Sch 1' (51.742N; 3.849E) in the deepest part of the lake (the Scharendijke basin) indicated by the black circle. The map was modified from Hagens et al., (2015).*

4.2.2 *Ex situ* experiments for AOM activity

The sediment slurry from 5-15 cm depth was homogenized in an anaerobic chamber and diluted with artificial seawater medium at a 1:2 ratio, and aliquots were placed in 120 ml sterile serum bottles and crimp sealed with rubber stoppers (40 ml headspace). The artificial seawater medium contained the following compounds per liter of demineralized water: NaCl (26 g), KCl (0.5 g) MgCl$_2$.6H$_2$O (5 g), NH$_4$Cl (0.3 g), CaCl$_2$.2H$_2$O (1.4 g), Na$_2$SO$_4$ (1.43 g), KH$_2$PO$_4$ (0.1 g), trace element solution (1 ml), 1 M NaHCO$_3$ (30 ml), vitamin solution (1 ml), thiamin solution (1 ml), vitamin B$_{12}$ solution (1 ml), 0.5 g L^{-1} resazurin solution (1 ml) and 0.5 M Na$_2$S solution (1 ml). The vitamins and trace element mixture was prepared according to Widdel and Bak (1992).

Quadrupled incubations were performed for each experiment containing seawater medium with sulfate as an electron acceptor. The headspace was filled with methane (99.9% CH$_4$ from Linde gas, Schiedam, the Netherlands) or nitrogen. These gases were pressurized to 2 bar in each incubation bottle. The bottles were incubated in the dark and shaken gently at room temperature (20-25°C). The batch incubations were performed for 250 days, liquid samples (2 ml) were taken every 15 days and headspace gas (2 ml) samples were collected every 50 days. The sampling was performed with a syringe equipped with a regulating valve to minimize potential pressure loss between sampling and analysis. The extracted sample volume was not replaced after sampling.

4.2.3 Analytical methods

Sulfate was measured by Ion Chromatography (Dionex-ICS-1000) after filtration with 0.45 μm filters as described previously (Sipma et al., 2004). Total dissolved sulfide (the sum of H$_2$S, HS$^-$, and S^{2-}) was determined by colorimetry using the methylene blue method (Acree et al., 1971). Samples were immediately injected into a 0.5 M NaOH solution to prevent volatilization of sulfide.

The volatile suspended solids (VSS), total suspended solids (TSS) and the dry weight of the sediment was estimated according to the procedure outlined in Standard Methods (APHA, 1995). Methane and carbon dioxide concentrations in the headspace of the incubations were measured by gas chromatography (GC 3800, VARIAN). The gas chromatograph was equipped with a PORABOND Q column (25 m × 0.53 mm × 10 μm) and a thermal conductivity detector. The carrier gas was helium (15 Psi), the oven temperature was 25°C and the gas valve injection was 0.5 ml. The measurements were performed in duplicate and the standard deviation of 0.5% of peak area was considered for calculation. For each sampling period, standard gas mixture of methane and carbon dioxide was injected to assure the quality measurement.

4.2.4 DNA extraction

DNA from the sediment (~ 0.5 g, the layer between 5 and 15 cm) obtained on December 2013, i.e., the same sediment used for the activity test, was extracted by the FastDNA® SPIN Kit for Soil (MP Biomedicals, Solon, OH, USA) following the manufacturer's protocol. The extracted DNA was quantified by a Nanodrop ND-1000 Spectrophotometer and the quality of DNA was checked by gel electrophoresis with 1% agarose.

4.2.5 PCR amplification of 16S rRNA genes and sequencing

The PCR for the sanger sequencing was performed by using the universal primer pairs for *Archaea* (Arc 21F and Arc 958R) and *Bacteria* (Eubac 27F and Eubac 1492R) as described by DeLong (1992). The PCR reaction mixture (50 μl) contained 2 μl of DNA template (~ 70 ng), 1X PCR buffer with 1.5 mM $MgCl_2$, 0.2 mM of deoxynucleoside triphosphates (dNTPs), 2.5 mg L^{-1} of Bovine serum albumin (BSA), 0.4 mM of each primer, 1.25 U of Ex Taq DNA polymerase (all Takara products, Japan) and nuclease free water. PCR amplification was performed with an Applied Biosystem Thermal Cycler with the following temperature program: i) initial denaturation step at 94°C for 5 min, ii) denaturation at 94°C for 40 min, iii) annealing at 55°C for *Archaea* and 50°C for *Bacteria* for 55 s, iv) elongation at 72°C for 40 s (30 cycles for *Archaea* and 25 cycles for *Bacteria*) and v) final elongation for 10 min. PCR products were purified using the E.Z.N.A.® Gel Extraction Kit by following the manufacturer's protocol (Omega Biotek, USA).

Archaeal and bacterial 16S rRNA genes from the variable region 4 (V4) were amplified for Illumina Miseq sequencing by using bar coded Arc519F and Arc806R primers as well as Bac520F and Bac802R primer sets for *Archaea* and *Bacteria*, respectively (Song et al., 2013). The PCR mixture and the conditions for Illumina Miseq was same as sanger sequencing, however 35 and 30 cycles were used for *Archaea* and *Bacteria* respectively and the annealing temperature for bacterial primer was 42°C.

4.2.6 Clone library and phylogenetic analysis

The amplified DNA was cloned with pMD18-T Simple Vector cloning kit (Takara, Japan) and then sequenced by a ABI 3730xl DNA analyzer (Applied Biosystems, USA). A total of 75 archaeal sequences and 60 bacterial sequences were obtained. The archaeal and bacterial 16S rRNA sequences retrieved from this study have been submitted to NCBI GenBank under the accession numbers KX088508-KX088617.

Sequences were checked with the Decipher chimera detection tool at http://decipher.cee.wisc.edu/FindChimeras.html, truncated sequences were excluded from the analysis. Reference sequences were retrieved using the BLAST network services (Altschul et al., 1997) and SILVA database (http://www.arb-silva.de) and were then aligned by using SINA web aligner (http://www.arb-silva.de/aligner/). Phylogenetic trees for *Archaea/Bacteria* specific sequences were constructed by using the Unweighted Pair Group Method with Arithmetic Mean (UPGMA) hierarchical clustering method (Sneath, 2005) as provided by MEGA 6 software (Tamura et al., 2013). The evolutionary distances of the phylogenetic tree was computed by the Kimura 2-parameter method (Kimura, 1980) as implemented by MEGA 6, with pair wise gap elimination, and 1000 bootstrap re-samplings for tree testing.

4.2.7 Illumina Miseq data processing

The purified DNA amplicons were sequenced by an Illumina HiSeq 2000 (Illumina, San Diego, CA, USA) generating 2 × 150 bp paired-end reads. A total of 48427 and 59267 sequences read for *Archaea* and *Bacteria* respectively from the DNA libraries passed the pipeline filters. After eliminating the chimeras, 48366 and 59216 sequences for *Archaea* and *Bacteria,* respectively, were analyzed and classified in MOTHUR (Schloss & Westcott, 2011). In short, the faulty sequences with mismatch tags or primers and with size less than 200 bp were removed by using the shhh.flows command. Then, the putative chimeric sequences were identified and removed by the chimera.uchime command using the most abundant reads in the respective sequence data sets as references. The sequence reads were classified according to the Silva taxonomy (Pruesse et al., 2007) using the classify.seqs command and the relative abundance of each phylotypes were estimated.

These sequence data have been submitted to the NCBI GenBank database under the BioProject PRJNA321334, under the accession KAFQ00000000 (direct link: http://www.ncbi.nlm.nih.gov/bioproject/PRJNA321334).

4.2.8 Fluorescence in situ hybridization (FISH) analysis

About 0.5 ml slurry was obtained from the batch bottles after 250 days of incubation, centrifuged and fixed with paraformaldehyde with a final concentration of 4% for 4 hours. The fixed sample was then stored at -20°C in a mixture of phosphate buffer saline solution (PBS) and ethanol (EtOH), with a PBS/EtOH ratio of 1:1 as described previously (Boetius et al., 2000). Stored samples were diluted with nuclease free water, sonicated for 40 s and then filtered on 0.2 μm white polycarbonate filters (GTTP type; 25 mm diameter; Millipore, Eschborn, Germany). The filtrate sample was hybridized with a mixture of Cy3-labeled ANME probes: ANME-1 350 (Boetius et al., 2000), ANME-2 538 (Treude et al., 2005a), ANME-3 1249 (Niemann et al., 2006b) and Cy5-labelled *DBB* 660 (Niemann et al., 2006b). Following the removal of unbound probes, cells were counterstained with 4',6-diamidino-2-phenylindole (DAPI) (Wagner et al., 1993). The hybridization and microscopic visualization of fluorescent cells was performed as described previously (Snaidr et al., 1997). Six independent FISH analysis were performed for each incubation test with methane and control without methane.

4.3 Results

4.3.1 Pore water chemistry of the Marine Lake Grevelingen sediment

Methane concentrations increased to around 6 mM below 20 cm depth in the sediment at the sampling site (Figure 4.2A). Formation of gas pockets and subsequent degassing observed during core-slicing suggest that part of the pore water methane might have been lost during sampling, thus potentially explaining the decreasing concentrations of methane with depth. The inorganic carbon concentration, i.e., $CaCO_3$, was around 2% between 5 and 20 cm depth, and showed an oscillating trend between 1.5 and 2.5% throughout the core (Figure 4.2B).

Similarly, the organic carbon content in the study site was about 3% at a depth of 5 and 20 cm and varied between 2.5 and 3.5% at depth below 20 cm (Figure 4.2C).

A gradual decrease in sulfate from 25 mM from the sediment surface to concentrations below 5 mM at a sediment depth of 35 cm was observed at the study site (Figure 4.2D). The highest sulfate removal rate in the sediment is estimated to have occurred around 10 cm depth. The concentration of total dissolved sulfide increased abruptly below 5 cm depth to nearly 6 mM at about 5 to 10 cm depth in November, whereas the maximum total dissolved sulfide concentration was observed at 10 to 20 cm depth in February (Figure 4.2E).

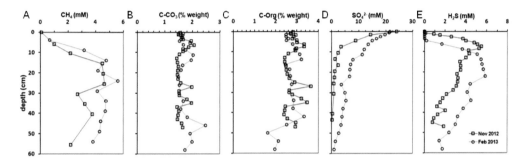

Figure 4.2 Depth profiles of key chemical pore water constituents and carbon in carbonates and organic matter in sediment of Marine Lake Grevelingen (A) CH_4, (B) $C\text{-}CaCO_3$, (C) C-Org, (D) SO_4^{2-} and (E) total dissolved sulfide (H_2S) for November 2012 and February 2013. Data are from Egger et al. (2016).

4.3.2 Ex-situ anaerobic oxidation of methane (AOM)

The incubation of sediment for 250 days with methane and sulfate as the sole supplied electron donor and electron acceptor, respectively, showed a gradual increment of total dissolved sulfide production and sulfate reduction (Figures 4.3A and 4.3B). The pH was around 7.5 at the start of the incubation and it increased up to 8.5 towards the end of the experiment.

The total dissolved sulfide concentration gradually increased throughout the experiment (Figure 4.3B). The total dissolved sulfide concentration in the batch bottles was around 6 mM at the end of the experiment, whilst almost 7.5 mM sulfate was consumed. After about 150 days of incubation, the total dissolved sulfide production was coupled to the consumption of sulfate at approximately equimolar ratios (Figures 4.3A and 4.3B). Minor differences in the expected production of sulfide based on the trend in sulfate consumption and the measured sulfide production are within the observational error based on the quadruplet incubations. Furthermore, sulfide can precipitate in the form of metal sulfides, especially as iron sulfide. The total dissolved sulfide and/or sulfate concentration remained almost constant in abiotic control incubations without biomass (Figs. 3A and 3B), indicating negligible abiotic sulfide production and/or sulfate consumption. In the control without methane, 2 mM of sulfate was consumed with the simultaneous production of sulfide, especially during the initial 100 days (Figures 4.3A and 4.3B) (Figs. 3A and 3B), suggesting that organoclastic sulfate reduction occurred as well. The maximum sulfate reduction rate from the incubations with methane was estimated at 5 µmol g_{dw}-1 d^{-1} (Table 4.1).

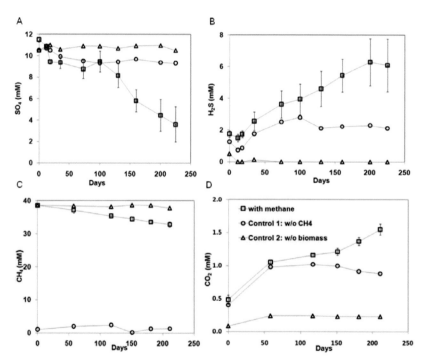

Figure 4.3 Concentrations of (A) dissolved sulfate and (B) total dissolved sulfide, and headspace (C) methane and (D) carbon dioxide during incubations of Marine Lake Grevelingen sediment with artificial sea water containing sulfate (10 mM) and headspace methane (2 bar). The error bar for the incubations with methane indicates the standard deviation (n=4).

The amount of methane and carbon dioxide in the headspace changed gradually during the period of incubation (Figures 4.3C and 4.3D). Parallel to the sulfate reduction, around 5.5 mM of methane was consumed during the incubation period (Figure 4.3C). The carbon dioxide concentration in the headspace increased over time (Figure 4.3D), however, 1.5 mM of carbon dioxide was detected in the batch tests at the end of the experiment. The control without sediment showed an almost constant methane concentration (38 mM), hence confirming the methane consumption only in the presence of sediment. Moreover, in the control incubation without methane, a small amount of methane was present already at the beginning, maybe entrapped in the sediment itself, and almost 2.4 mM of methane was produced during the first 100 days of incubation, whereas the concentration of carbon dioxide in the control experiments was also low (Figure 4.3D).

4.3.3 Relative abundance of archaeal and bacterial population

A total of 50,000 (± 10,000) sequence reads were obtained for each bacterial and archaeal 16S rRNA gene amplicon from the investigated sediment layer via high-throughput sequencing. The majority of archaeal sequences were affiliated with uncultured phylotypes within the archaeal phyla belonging to the *Euryarchaeota* (43%), *Thaumarchaeota* (38%) and *Crenarchaeota* (3.3%), whereas 15% of the archaeal sequences could not be closely affiliated with any known phylum and are thus referred to as unclassified *Archaea* (15%). The *Euryarcharchaeota* was the archaeal phylum with a higher relative abundance in the Marine Lake Grevelingen sediment, which also includes the archaeal phylotypes with a major role in the methane cycle. The majority of the sequences clustering with the Euryarchaeotic phyla were affiliated to the uncultured *Euryarchaeota* (Fig. 4A). *Methanomicrobia*, a methanogenic clade, were also retrieved from the sediment, their relative abundance was, however, less than 1% (Fig. 4A). Putative ANME-3 affiliated sequences were also retrieved with a relative abundance of ~ 8.5%.

Other diverse archaeal clades with diverse metabolic functions were also present in the retrieved sequences. Within the *Thaumarchaeota* phylum, the *Nitrosopumilales* and *Nitrososphaerales* class were observed (Fig. 4A). The relative abundance of the *Nitrosopumilales* (38%) was much higher than the relative abundance of the

Nitrososphaerales (0.5%). Only 3.3% of the total archaeal sequences were affiliated to the *Thermoprotei* class of the *Crenarchaeota* phylum (Fig. 4A).

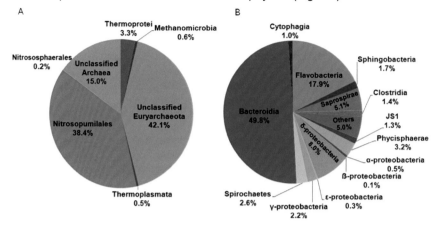

Figure 4.4 Archaeal (A) and bacterial (B) community composition in the Marine Lake Grevelingen sediment at the class level as determined by high throughput sequencing. The relative abundance (% of 16S rRNA gene sequences) of each phylotype is given.

The bacterial high throughput sequence analysis (Figure 4.4B) showed a diverse bacterial community with almost 18 classes, distributed among different phyla including *Bacteroidetes*, *Firmicutes*, *Planctomycetes*, *Proteobacteria* and *Spirochaetes*. The most abundant bacterial class was *Bacteroidia* with almost 50% sequences affiliated to it, followed by *Flavobacteria*, *Deltaproteobacteria* and *Saprospirae* (Figure 4.4B). The *Proteobacteria* phylum was diversified with *Alphaproteobacteria*, *Betaproteobacteria*, *Gammaproteobacteria*, *Deltaproteobacteria* and *Epsilonproteobacteria* (Figure 4.4B). However, the relative abundance of other *Proteobacteria* was less than 1% except for *Deltaproteobacteria* and *Gammaproteobacteria* (Figure 4.4B). The relative abundance of *Deltaproteobacteria* phyla, which incorporates the sulfate reducing bacteria, was about 8% of the total bacterial sequences retrieved.

4.3.4 Archaeal and bacterial phylogeny

The microbial community in the sediment layer from the study site was further investigated through phylogenetic analysis of 70 archaeal and 56 bacterial clone sequences (Figures 4.5 and 4.6). Distinct clusters from seven different phylogenetic lineages were observed for *Archaea* (Figure 4.5). In agreement with high throughput sequences analysis, the majority of the retrieved sequences belonged to the *Thaumarchaeota* and *Euryarchaeota* phyla. The first two clusters of clones in the archaeal phylogenetic tree (almost 42% of total sequences) were most closely related to the archaeal sequences of uncultured environmental *Thaumarcheaota* sequences with the majority of them from the *Nitrosopumilales* order, as shown by the same phylogenetic lineage with *Candidatus Nitrospumilus sp.* (Figure 4.5).

The second largest archaeal cluster (with almost 30% of the total clones sequenced) was the uncultured *Euryarchaeota* group which includes methanogens, such as *Methanomicrobiales* and *Methanosarcinales* related clones and also ANME. Among all archaeal sequences, 7 clones (ᶠ 10% of total sequences analyzed) were closely affiliated to uncultured archaeal sequences related to ANME-3 from the Haakon Mosby mud volcano (identification level: 99%). This is the thus far only reported ANME habitat with a dominance of ANME-3 (Niemann et al., 2006b). Besides ANME-3, five euryarchaeotal clones were affiliated to uncultured *Methanosarcinales,* among which two of them were closely associated with *Methanosaeta sp.* and others with *Methanosarcina sp.* However, none of the retrieved sequences could be linked with the ANME-1 or ANME-2 clades of anaerobic methanotrophs by the method used.

Other clusters of Marine Lake Grevelingen sequences were associated to environmental sequences from hydrothermal vents and clustered within the Marine Hydrothermal Vent Group, Marine Benthic Group (B and D) and 6 clones were putatively affiliated with the Miscellaneous *Crenarchaeota* Group (Figure 4.5), which has recently been named as *Bathyarchaeota* (Attar, 2015).

The bacterial 16S rRNA sequences were mainly distributed among the phyla *Proteobacteria, Spirochaete, Bacteroidetes* and *Planctomycetes* (Figure 4.6A). The majority of the sequences (ᶠ 35%) were affiliated to the phylum *Proteobacteria,* including the classes *Alphaproteobacteria, Gammaproteobacteria, Deltaproteobacteria* and *Epsilonproteobacteria*. Most of the *Proteobacteria* from the studied site were closely related to the *Proteobacteria* previously retrieved from marine sediments (Figure 4.6).

The first two clusters in the bacterial phylogenetic tree were *Gammaproteobacteria* and *Alphaproteobacteria*. Within the *Gammaproteobacteria* (5 clones), only one clone sequence GV bac 3 (Figure 4.6C) was closely related to a previously retrieved Lake Grevelingen clone (Vasquez-Cardenas et al., 2015). The rest of the retrieved gammaproteobacterial clones were affiliated to uncultured *Proteobacteria* from other coastal sediments. Furthermore, only three sequences were affiliated with the *Alphaproteobacteria*, among which only one clone could be linked to a known strain of *Pseudorhodobacter sp.*

A total of six clones, i.e., almost 10% of the total bacterial sequences, were affiliated with different sulfate reducing bacteria belonging to the *Deltaproteobacteria* among which only one clone was putatively linked with *Desulfobulbaceae* (Figure 4.6B). The retrieved *Desulfobulbaceae* sequence GV bac 14 was clustered together with the previously retrieved *Desulfobulbaceae* clone from the sediment sampled at 30 m water depth of Lake Grevelingen (Vasquez-Cardenas et al., 2015). Furthermore, one clone of *Desulfobacter hydrogenophilus*, a type of hydrogen utilizing sulfate reducing bacteria was retrieved. Some of the retrieved *Deltaproteobacteria* were clustered together with the cultured strain of *Desulfobacterium*, whereas GV bac 27 was associated with the sulfate reducing bacteria previously retrieved from Eel river (Eel-1 group) (Beal et al., 2009). Moreover, one clone (GV bac 54) was affiliated with *Sulfurimonas* sp. of the *Epsilonproteobacteria* and was phylogetically similar to the clone previously retrieved from the other sites of Marine Lake Grevelingen (Vasquez-Cardenas et al., 2015).

In agreement with the high throughput sequence results, a large cluster (8 clones) belonged to the class *Bacteroidetes*. One clone of *Bacteroidetes* was affiliated with the *Bacterioidetes* previously found in polluted coastal sediment, clustered together with *Lewinella sp.* and belonged to the order of *Spingobacteria* (Acosta-González et al., 2013). Other retrieved *Bacteroidetes* sequences formed a clear cluster and distributed among the families *Flavobacteriaceae* (*Maritimimonas sp.* and *Sufflavibacter sp.*), *Bacteroidales* (*Marinifilum sp.*) and *Cytophagia* (Figure 4.6A). Besides the clade of *Proteobacteria* and *Bacteroidia*, the bacterial phylogeny of Marine Lake Grevelingen sediment was diversified with clones related to different uncultured marine bacteria related to *Spirochaetes*, *Planctomycetes*, and a few clones of *Acidobacteria*, *Clostridia*, *Verrumicrobia* and *Choloroflexi*.

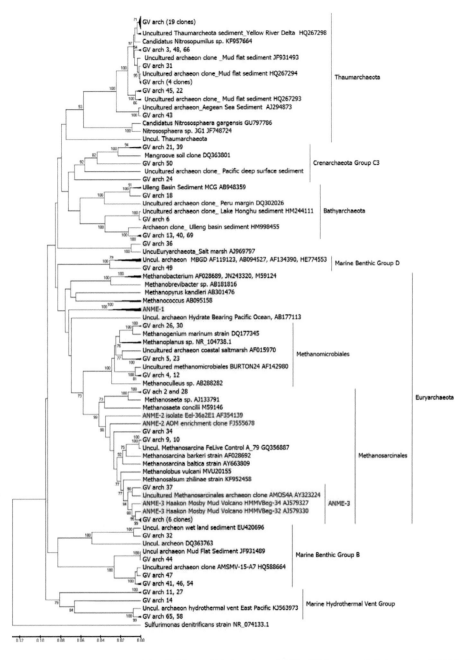

Figure 4.5 *Phylogenetic tree of the PCR-amplified archaeal 16S rRNA gene phylotypes (c. 1000 bp) in the sediments of Marine Lake Grevelingen, based on UPGMA as determined by distance using Kimura's two-parameter method as implemented by MEGA 6 (Tamura et al., 2013). Each phylotype recovered is named after GV arch and respective clone numbers. Numbers at base of nodes are bootstrap proportions larger than 70% (1000 replicates). Scale bar represents substitutions per unit distance per site.*

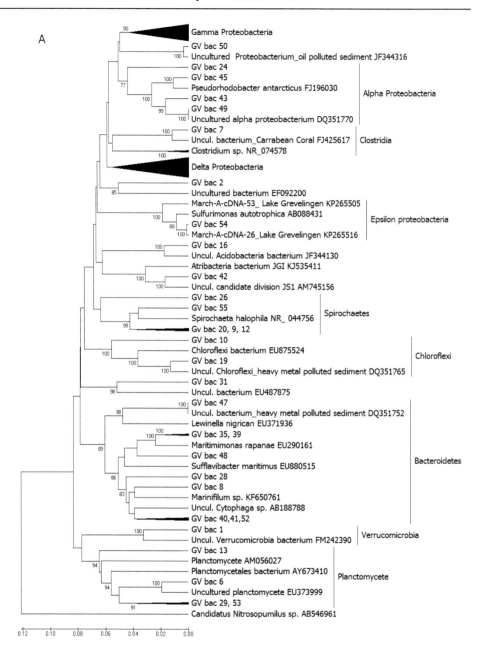

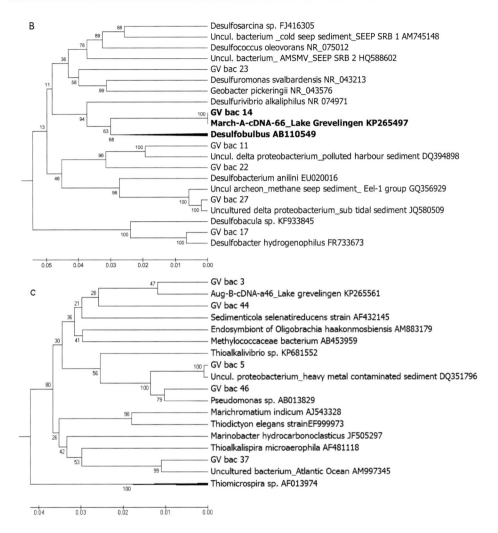

Figure 4.6. *(A) Phylogenetic tree of the PCR-amplified bacterial 16S rRNA gene phylotypes (c. 1400 bp) in the sediments of Marine Lake Grevelingen and partial tree showing the details of collapsed branches for (B) Deltaproteobacteria and (C) Gammaproteobacteria, based on UPGMA as determined by distance using Kimura's two-parameter method as implemented by MEGA 6 (Tamura et al., 2013). Each phylotype recovered is named after GV bac and respective clone numbers. The reference clones of Marine Lake Grevelingen were retrieved from a previous study from the hypoxic location of Lake Grevelingen by Vasquez-Cardenas et al. (2015). Numbers at base of nodes are bootstrap proportions larger than 70% (1000 replicates). Scale bar represents substitutions per unit distance per site.*

4.3.5 Visualization of archaeal and bacterial cells

Archaeal cells were visualized by FISH analysis after 250 days of incubations both in the activity test incubation with methane and the control without methane (Figures 4.7A and 4.7E). Moreover, the archaeal cells were clearly visualized (almost 30 - 40% of the total cells) in the samples incubated with methane, whereas they were less abundant (~ 5-10% of total cells) in most of the cases for the control without methane. The cocci shaped ANME cells (Cy3-labelled ANME-1 350, ANME-2 538, ANME-3 1249) in the incubation with methane were observed as single type cells with each cells of 1.7 µm diameter in the incubation with methane (Figure 4.7C). We conclude that these cells are likely ANME-3 since no other ANME types were detected by phylogenetic analysis. In the same sample of incubation with methane, cocci-shaped *Desulfobulbaceae* cells with almost 1.5 µm in diameter were observed, however, no ANME cells were found in their vicinity (Figure 4.7B). On the contrary, in the samples from control incubation without methane, ANME cells could not be visualized even after several attempts of hybridization (Figure 4.7F). Similarly, the *Desulfobulbaceae* cells were not observed in the control incubation without methane (images not presented here).

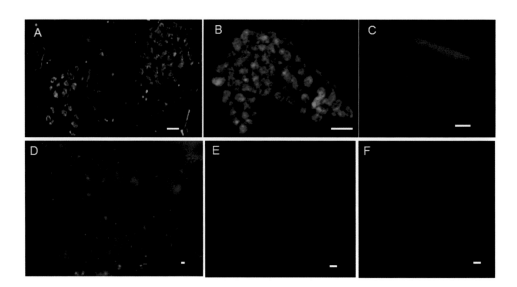

Figure 4.7 FISH images of microbial cells: (A-C) photomicrographs obtained at the end of incubation with methane and sulfate and (D-F) photomicrographs obtained at the end of control incubation without methane. Photomicrographs of DAPI stained cells are in blue color (A and D), archaeal cells

hybridized by Arc-FAM probe are in red (A and E), ANME-2 hybridized with ANME-mix probe are in red (F and F), and Desulfobulbaceae hybridised with DBB 660 are in green color (B). Scale bar refers to 2 µm.

4.4 Discussion

4.4.1 AOM occurrence and activity

This study showed potential AOM coupled to sulfate reduction in activity tests with Marine Lake Grevelingen sediment after 100 days of incubation as evidenced by sulfate reduction, sulfide production and simultaneous methane consumption. Net AOM dependent sulfate reduction did not occur during the first 100 days, after which the sulfate reducing activity of the incubations with and without methane diverged. Over the first 100 days, residual organic matter likely provided the electrons for sulfate reduction, and only when organic matter dependent sulfate reduction decreased, AOM induced sulfate reduction became active. Similarly, for the high pressure incubation with Eckernförde Bay sediment, the carbon dioxide production from methane was faster when the endogenous organic content became depleted (Timmers et al., 2015). Trace methane oxidation can occur even when net methane production is observed (Meulepas et al., 2010; Timmers et al., 2015), also in conditions in which the carbon dioxide and sulfide production rises. However, in the control incubation without methane, only little methane was produced in the first 100 days, after which methane concentrations remained stable (Figure 4.3). Although trace methane oxidation cannot be excluded from this set of experiments, AOM coupled to sulfate reduction became the dominant process after prolonged incubation.

The *ex situ* AOM induced sulfate reduction rate by the Marine Lake Grevelingen sediment lies within the range of sulfate reduction rates observed in other marine sediments (Table 4.1). *In situ* sulfate reduction rates in the upper 15 cm of sediment (50-170 nmol cm^{-3} d^{-1} or ~ 0.16-0.53 µmol cm^{-3} d^{-1} or 0.05-0.17 µmol g_{dw}^{-1} d^{-1}) (Egger et al., 2016) were lower than *ex situ* rates but also fall within the range observed in other systems. Previous work at the study site demonstrated that methanogenesis and sulfate reduction occur simultaneously in the surface sediments due to the high input of organic matter (Egger et al., 2016). The authors further hypothesized that

AOM is occurring, but that rates of AOM are likely too low to efficiently remove pore water methane due to the high sediment accumulation rate and the slow growth of methanotrophic communities. The *in vitro* incubation with methane in this study showed a linear decrease in methane throughout the incubation period supporting the previous conclusion of active AOM in the sediments of the Scharendijke basin of Marine Lake Grevelingen.

After 200 days, the concentration of total dissolved sulfide did not increase further. If this trend is real (note the large error bars, Figure 4.3B), this may be caused by inhibition of microbial activity due to sulfide toxicity. Indeed, some studies report the inhibition of AOM dependent sulfate reduction activity by sulfide toxicity during batch incubation, for instance at concentrations of ~ 2.5 mM for Eckernförde Bay sediment (Meulepas et al., 2009) and ~ 4 mM for sediment from the Captain Aryutinov mud volcano of the Gulf of Cadiz at 10 MPa (Zhang et al., 2010). However, ANME-3, present in Lake Grevelingen could be less prone to sulfide toxicity since it normally occurs in environments with high sulfide concentrations (Niemann et al., 2006).

The amount of methane consumed in the headspace was generally in accordance with the amount of total dissolved sulfide produced/ sulfate reduced (Figures 4.3C and 4.3D) and followed the expected 1:1 stoichiometry of sulfate-dependent AOM. In the control without supplied methane, methanogenic and sulfate reducing activity was observed at the start of the incubations, which can be attributed to the degradation of endogenous organic matter (Figure 4.3). This likely also explains the abrupt increase in carbon dioxide during the starting of the incubation. Moreover, for the control without added methane, methanogenesis and trace methane oxidation might also contribute to the production of carbon dioxide. Carbon dioxide concentration in the headspace was almost three times lower than the amount of methane oxidized which was mainly attributed by the change of pH from 7.5 to 8.5 in the activity batches resulting in the large amount of carbonate species rather than headspace carbon dioxide. In the natural environment also, most of the microbially produced inorganic carbon precipitates as carbonate minerals and dolomite formation occurs (Zhang, 2011). To some extent, the produced inorganic carbon might be utilized by the ANME and/or other microorganisms in the sediment. Carbon dioxide assimilation by ANME populations was suggested previously for several marine sediments. Examples include the consumption of carbon dioxide by ANME-1

128

from Black Sea microbial mats (Treude et al., 2007) and the autotrophic mode of AOM by ANME-1 from the hydrothermal vent Guaymas basin (Kellermann et al., 2012).

4.4.2 ANME and sulfate reducing bacteria in the Marine Lake Grevelingen sediment

In the previous geochemical study of the Scharendijke basin of Marine Lake Grevelingen, there were indications of AOM although at low rates ($\sim$ 0.05-0.17 µmol g_{dw}^{-1} d^{-1}) (Egger et al., 2016). This study verifies the presence of ANME, and more specifically, ANME-3 in the studied sediment. Along with methane consuming ANME, the Marine Lake Grevelingen sediment contains archaeal clades responsible for methanogenesis, suggesting that the microbial ecosystem in the sediment is involved in methane metabolism.

The retrieval of ANME-3 clones from Marine Lake Grevelingen sediment contrasts with previous observations of only ANME-1 and ANME-2 in coastal sediments. Examples include observations of ANME-2 in Eckernförde Bay sediment (Treude et al., 2005b), ANME-1 in Aarhus Bay sediment (Aquilina et al., 2010) and ANME-2 in Skagerrak sediment, Denmark (Parkes et al., 2007) (Table 4.1). ANME-3 are typically found only in cold seep areas and mud volcanoes with high methane partial pressures and temperatures < 20 °C. The major habitats of ANME-3 include Haakon Mosby mud volcano at a water depth of 1250 m (Niemann et al., 2006b), Kazan mud volcano in the Eastern Mediterranean at a water depth of 1700 m (Heijs et al., 2007; Pachiadaki et al., 2010), and surface sediment from Guaymas basin at a water depth of 1500 m (Vigneron et al., 2013a).

As shown by Egger et al. (2016) methane fluxes at this study site are three orders of magnitude higher than is common in diffusive sediments and fall within the low range of methane fluxes reported for marine seep sediments. This study thus indicates that ANME-3 can also occur in coastal sediments with high methane fluxes, albeit with high rates of sediment accumulation, thereby extending the previously known environments, i.e., seeps and mud volcanoes. With the finding of ANME-3 present in sediment at a water depth of only 45 m (i.e. 0.45 MPa pressure), this study also shows that the ANME-3 can be enriched at atmospheric pressures and may also tolerate rather high sulfide concentrations, which is advantageous for further physiological investigation (Gonzalez-Gil et al., 2011).

The sulfate reduction process was determined by both *in situ* (Egger et al., 2016) and *in vitro* methods. *In vitro* sulfate reduction and total dissolved sulfide production rates were in the range of other marine sediments (Table 1). However, the sequence analysis and clone library of the bacterial community showed the relative abundance and diversity of sulfate reducing bacteria related *Deltaproteobacteria* is low at this study site. It should be noted, however, that the diversity analysis was performed at the DNA level, which was not a direct indicator of the metabolic active players. A metagenome study on AOM enrichments described the dominance of sulfate reducing bacteria at the transcriptomic level (Wang et al., 2013). However, only few clones related to sulfate reducing bacteria were retrieved from gene-based analysis indicating relatively lower abundance. Among the common sulfate reducing bacteria involved in AOM, only one clone of *Desulfobulbaceae* was found. Other clades of sulfate reducing bacteria, but only few clones, were also retrieved such as *Desulfobacter* (Eel-1), *Desulfuromonas* and other uncultured *Deltaproteobacteria* which are not yet known for an involvement in AOM. However, these clades play an important role in sulfate reduction and in sulfur cycling especially in marine environments and can use a wide range of carbon sources such as acetate, hydrogen and other volatile fatty acids (Muyzer & Stams, 2008). Furthermore, the *Desulfosarcina* / *Desulfococcus* clades of sulfate reducing bacteria commonly retrieved in ANME habitats dominated by ANME-1 and ANME-2 were not found at the site studied, suggesting that *Desulfobulbaceae* are the only sulfate reducing bacteria present at the study site, which is a potential partner of ANME-3.

Desulfobulbaceae have been found in ANME habitats where mostly ANME-3 were present as well as at some other sites, such as the Sonara margin cold seep (Vigneron et al., 2013a; Vigneron et al., 2013b), Haakon Mosby mud volcano with a majority of ANME-3 (Losekann et al., 2007) and the Eel river with dominance of ANME-1 and ANME-2, suggesting *Desulfobulbaceae* have a role in AOM (Orphan et al., 2001). Similar to this study only one *Desulfobulbaceae* clone was retrieved from Haakon Mosby mud volcano and also only a few *Desulfobulbaceae* sequences were retrieved from the Eel river and Sonara margin, so very few environmental sequences of this clade have been obtained so far (Losekann et al., 2007).

130

Table 4.1. Comparison of sulfate, methane and AOM induced sulfate reduction rate (in vitro) of marine sediments.

Location	Water depth (m)	CH$_4$* (mM)	SO$_4^{2-}$* (mM)	Sulfide* (mM)	Major ANME types	SR (µmol g$_{dw}^{-1}$ d^{-1})	Reference
Black Sea (giant carbonate chimney)	230	2.8	17		ANME-1	4.3 - 19	(Michaelis et al., 2002; Treude et al., 2007)
Guaymas Basin hydrothermal vent	-	-	-		ANME-1	0.25	(Holler et al., 2011)
Juan de Fuca Ridge hydrothermal vent	2400	3	-		ANME-1	-	(Wankel et al., 2012)
Black Sea (other microbial mats)	180	3.7	9 to 15	2	ANME-1 and ANME-2	4-20	(Krüger et al., 2008)
Gulf of Cadiz, 286 d pressure bioreactor enrichment	900	37			ANME-2	9.22	(Zhang et al., 2010)
Eckernförde Bay sediment	28	0.2 to 0.8	20		ANME-2	0.5 µmol cm^{-3} d^{-1}	(Treude et al., 2005b)
Subsurface marine sediments, Skagerrak	308	2	20		ANME-2	-	(Parkes et al., 2007)
Marine Lake Grevelingen sediment	**40**	**6**	**25**	**6**	**ANME-3**	**5**	**This study**
Haakon Mosby mud volcano	1250	0.0058	28	4	ANME-3	-	(Losekann et al., 2007)

Note: * refers to these data are based on *in situ* measurement

Single shell type ANME cells and *Desulfobulbaceae* cells were observed during FISH analysis, and thus not as a tight co-aggregation of ANME and sulfate reducing bacteria as previously observed in various marine sediments with ANME-1 and ANME-2 (Boetius et al., 2000; Orphan et al., 2001; Vigneron et al., 2013a). It is unlikely that the ANME-3/ *Desulfobulbaceae* aggregates were destroyed during sample preparation for FISH, as the samples were withdrawn and fixed without manipulation. Although, consortia of ANME-3 and *Desulfobulbaceae* were dominant in Haakon Mosby mud volcano sediment, also abundant single cells and loose aggregates of ANME-3 have been observed in this sediment (Losekann et al., 2007) . The *Desulfobulbaceae* and loosely associated ANME-3 cells were present at a ratio lower than one. This suggests that the association with *Desulfobulbaceae* is not and obligatory partner and ANME-3 archaea may not need a sulfate reducing partner for AOM.

Recent studies reported different interaction modes between ANME and their bacterial partner. Milucka et al. (2012) postulated that a syntrophic partner is generally not needed for ANME and that they are possibly responsible for both AOM and sulfate. Other recent studies pointed towards cooperative interactions between ANME and sulfate reducing bacteria based on interspecies electron transfer, both for thermophilic ANME-1 and mesophilic ANME-2 consortia (McGlynn et al., 2015; Wagner, 2015). However, Scheller et al. (2016) showed that ANME-2 and sulfate reducing bacteria can be decoupled in the laboratory by providing different soluble electron acceptors, such as ferric iron. Therefore, ANME-2 seems capable of respiratory metabolism, which could be the reason why ANME cells are frequently found alone without the bacterial partner. It is possible that ANME can also use insoluble ferric oxides when they are abundant in the sediment. The coupling or decoupling of ANME from the bacterial partner thus possibly depends on the environmental conditions and the ANME type. Contents of reducible Fe oxides in the sediment at this site were > 50 μmol g^{-1} (Egger et al., 2016), indicating a potential for AOM coupled to Fe-oxide reduction. An enrichment of ANME-3 would facilitate further investigations of ANME-3 metabolism and their interactions with sulfate reducing bacteria. One of the interesting features of Lake Grevelingen is the occurrence of a novel cable bacterium in the lake sediments (Vasquez-Cardenas et al., 2015). These cable bacteria are putatively linked to the group of

Desulfobulbaceae, and are known to be sulfur oxidizers, however whether the cable bacteria with the electrogenic filaments have any involvement with or facilitate electron transfer in AOM has to be investigated in future.

4.4.3 Diversity of other archaeal and bacterial clades

One of the larger clades of *Archaea* retrieved from the Marine Lake Grevelingen sediment was the *Nitrosopumilus* clade of *Thaumarcheota*. The cultured strains of this clade are known as ammonia oxidizers and assimilate organic compounds (Qin et al., 2014). These microbial clades thus potentially play a role in carbon and nitrogen cycling in the marine sediments. Some clades of *Thaumarcheoata*, also known as marine group I, were retrieved from different marine environments, namely Aarhus Bay sediment (Aquilina et al., 2010) and other deep marine hydrate and seeps sediments (Inagaki et al., 2006; Vigneron et al., 2013a).

Other groups of *Archaea* found in the studied sediment are the *Bathyarchaeota* and the Marine Benthic Group (MBG), which are similar to the clade found in previous studies from marine sediments and cold seeps. The *Barthyarchaeota* are widely distributed throughout marine sediments and have also been found in some terrestrial environments (Lloyd et al., 2013). Their ecophysiological role has not yet been well defined, genomic data have suggested that *Barthyarchaeota* could degrade buried organic carbon (Biddle et al., 2006), especially aromatic compounds, to produce acetate or even methane (He et al., 2016; Lazar et al., 2016; Meng et al., 2014).

Similar to sulfate reducing bacteria, other bacterial clades with a role in sulfur metabolism were also observed, even with low relative abundance. Seasonal dominancy of *Beggiatoa* mats with sulfur oxidizers such as *Thiomicrospira sp.* and *Thioalkaispira sp.* is common at different sites in the Lake Grevelingen (Seitaj et al., 2015). However, only one bacterial clone (GV bac 37) was distantly affiliated with the *Thioalkaispira sp.,* suggesting the investigated sample from the potential SMTZ of the sediment layer may lie below the sulfide oxidizing mat. Moreover, the bacterial sequences related to aerobic methanotrophs from *Gammaproteobacteria*, such as *Methylobacter* and *Methylomonas* groups, were not found in the sediment by the

133

method used, supporting anaerobic methanotrophy is main mode of microbial methanotrophy in the investigated site.

Bacteroidetes were one of the predominant *Bacteria* at the study side. Past studies have shown only a few percentages of *Bacteroidetes* or even no *Bacteroidetes* in the bacterial community at sites with AOM and active sulfate reduction, for example in the bacterial community of Sonara margin cold seep (Vigneron et al., 2013b), Aarhus bay (Aquilina et al., 2010) and the Ulleng basin (Lee et al., 2013) sediment. *Bacteroides* are obligate anaerobes and their dominance suggests the investigated sediment is completely anaerobic, further supporting anaerobic methanotrophy is the methane oxidizing pathway. Another major class of *Bacteroidetes*, i.e *Flavobacteria*, was also found (Figures 4.4B, and 4.6A). These *Bacteria* are only recovered from the marine water column (Brettar et al., 2004; Labrenz et al., 2007). We hypothesized that the high rates of sediment accumulation at this site contributed to rapid burial of *Flavobacteria* cells/DNA in the sediment, thereby explaining their occurrence below the sediment-water interface.

4.5 Conclusions

This study demonstrates AOM activity by *in vitro* incubations and *in situ* geochemical profile studies of coastal sediment from Marine Lake Grevelingen. ANME-3 type anaerobic methanotrophs were retrieved at the studied site, whereas other ANME clades, i.e. ANME-1 and ANME-2 were not detected in this study. Among the common sulfate reducing bacteria associated to ANME, *Desulfobulbaceae* cells were observed and only few sequences were linked with sulfate reducing bacteria clades.

4.6 References

Acosta-González, A., Rosselló-Móra, R., Marqués, S. 2013. Characterization of the anaerobic microbial community in oil-polluted subtidal sediments: aromatic biodegradation potential after the Prestige oil spill. *Environ. Microbiol.*, **15**(1), 77-92.

Acree, T.E., Sonoff, E.P., Splittstoesser, D.F. 1971. Determination of hydrogen sulfide in fermentation broths containing SO_2. *Appl. Environ. Microbiol.*, **22**(1), 110-112.

Altschul, S., Madden, T., Schaffer, A., Zhang, J., Zhang, Z., Miller, W., Lipman, D. 1997. Gapped BLAST and PSI-BLAST: a new generation of protein database search programs. *Nucleic Acids Res.*, **25**(17), 3389-3402.

APHA. 1995. Standard methods for the examination of water and wastewater, *American Public Health Association.* (19th edition), Washington DC, USA: p 1325

Aquilina, A., Knab, N.J., Knittel, K., Kaur, G., Geissler, A., Kelly, S.P., Fossing, H., Boot, C.S., Parkes, R.J., Mills, R.A., Boetius, A., Lloyd, J.R., Pancost, R.D. 2010. Biomarker indicators for anaerobic oxidizers of methane in brackish-marine sediments with diffusive methane fluxes. *Org. Geochem.*, **41**(4), 414-426.

Attar, N. 2015. Archaeal genomics: A new phylum for methanogens. *Nature Reviews Microbiology*, **13**(12), 739-739.

Beal, E.J., House, C.H., Orphan, V.J. 2009. Manganese- and iron-dependent marine methane oxidation. *Science*, **325**(5937), 184-187.

Biddle, J.F., Cardman, Z., Mendlovitz, H., Albert, D.B., Lloyd, K.G., Boetius, A., Teske, A. 2012. Anaerobic oxidation of methane at different temperature regimes in Guaymas Basin hydrothermal sediments. *ISME J.*, **6**(5), 1018-1031.

Biddle, J.F., Lipp, J.S., Lever, M.A., Lloyd, K.G., Sorensen, K.B., Anderson, R., Fredricks, H.F., Elvert, M., Kelly, T.J., Schrag, D.P., Sogin, M.L., Brenchley, J.E., Teske, A., House, C.H., Hinrichs, K.-U. 2006. Heterotrophic Archaea dominate sedimentary subsurface ecosystems off Peru. *Proc. Natl. Acad. Sci. USA*, **103**(10), 3846-3851.

Boetius, A., Holler, T., Knittel, K., Felden, J., Wenzhöfer, F. 2009. The seabed as natural laboratory: lessons from uncultivated methanotrophs. in: Epstein, S.S. (Ed.), *Uncultivated Microorganisms*. Vol 10 of series Microbiology monographs, Springer Berlin Heidelberg, Germany, pp. 293-316.

Boetius, A., Ravenschlag, K., Schubert, C.J., Rickert, D., Widdel, F., Gieseke, A., Amann, R., Jorgensen, B.B., Witte, U., Pfannkuche, O. 2000. A marine microbial consortium apparently mediating anaerobic oxidation of methane. *Nature*, **407**(6804), 623-626.

Boetius, A., Wenzhöfer, F. 2013. Seafloor oxygen consumption fuelled by methane from cold seeps. *Nat. Geosci.*, **6**(9), 725-734.

Brazelton, W.J., Schrenk, M.O., Kelley, D.S., Baross, J.A. 2006. Methane - and sulfur-metabolizing microbial communities dominate the Lost City hydrothermal field ecosystem. *Appl. Environ. Microbiol.*, **72**(9), 6257-6270.

Brettar, I., Christen, R., Höfle, M.G. 2004. *Belliella baltica* gen. nov., sp. nov., a novel marine bacterium of the *Cytophaga–Flavobacterium–Bacteroides* group isolated from surface water of the central Baltic Sea. *Int. J. Syst. Evol. Micr.*, **54**(1), 65-70.

Dale, A.W., Aguilera, D., Regnier, P., Fossing, H., Knab, N., Jørgensen, B.B. 2008. Seasonal dynamics of the depth and rate of anaerobic oxidation of methane in Aarhus Bay (Denmark) sediments. *J. Mar. Res.*, **66**(1), 127-155.

DeLong, E.F. 1992. Archaea in coastal marine environments. *Proc. Natl. Acad. Sci. USA*, **89**(12), 5685-5689.

Egger, M., Lenstra, W., Jong, D., Meysman, F.J., Sapart, C.J., van der Veen, C., Röckmann, T., Gonzalez, S., Slomp, C.P. 2016. Rapid sediment accumulation results in high methane effluxes from coastal sediments. *PloS one*, **11**(8), e0161609.

Ettwig, K.F., Butler, M.K., Le Paslier, D., Pelletier, E., Mangenot, S., Kuypers, M.M.M., Schreiber, F., Dutilh, B.E., Zedelius, J., de Beer, D., Gloerich, J., Wessels, H.J.C.T., van Alen, T., Luesken, F., Wu, M.L., van de Pas-Schoonen, K.T., Op den Camp, H.J.M., Janssen-Megens, E.M., Francoijs, K.-J., Stunnenberg, H., Weissenbach, J., Jetten, M.S.M., Strous, M. 2010. Nitrite-driven anaerobic methane oxidation by oxygenic bacteria. *Nature*, **464**(7288), 543-548.

Gonzalez-Gil, G., Meulepas, R.J.W., Lens, P.N.L. 2011. Biotechnological aspects of the use of methane as electron donor for sulfate reduction. in: Murray, M.-Y. (Ed.),*Comprehensive Biotechnology*. Vol. 6. (2nd edition), Elsevier B.V., Amsterdam, the Netherlands pp. 419-434.

Hagens, M., Slomp, C.P., Meysman, F.J.R., Seitaj, D., Harlay, J., Borges, A.V., Middelburg, J.J. 2015. Biogeochemical processes and buffering capacity concurrently affect acidification in a seasonally hypoxic coastal marine basin. *Biogeosciences*, **12**(5), 1561-1583.

Haroon, M.F., Hu, S., Shi, Y., Imelfort, M., Keller, J., Hugenholtz, P., Yuan, Z., Tyson, G.W. 2013. Anaerobic oxidation of methane coupled to nitrate reduction in a novel archaeal lineage. *Nature*, **500**(7464), 567-570.

He, Y., Li, M., Perumal, V., Feng, X., Fang, J., Xie, J., Sievert, S., Wang, F. 2016. Genomic and enzymatic evidence for acetogenesis among multiple lineages of the archaeal phylum Bathyarchaeota widespread in marine sediments. *Nat. Microbiol.*, **1**(16035), 1-9.

Heijs, S.K., Haese, R.R., van der Wielen, P.W., Forney, L.J., van Elsas, J.D. 2007. Use of 16S rRNA gene based clone libraries to assess microbial communities potentially involved in anaerobic methane oxidation in a Mediterranean cold seep. *Microb. Ecol.*, **53**(3), 384-398.

Hinrichs, K.-U., Hayes, J.M., Sylva, S.P., Brewer, P.G., DeLong, E.F. 1999. Methane-consuming archaebacteria in marine sediments. *Nature*, **398**(6730), 802-805.

Hinrichs, K.-U., Summons, R.E., Orphan, V.J., Sylva, S.P., Hayes, J.M. 2000. Molecular and isotopic analysis of anaerobic methane-oxidizing communities in marine sediments. *Org. Geochem.*, **31**(12), 1685-1701.

Hoehler, T.M., Alperin, M.J., Albert, D.B., Martens, S. 1994. Field and laboratory studies of methane oxidation in an anoxic marine sediment: Evidence for a methanogen-sulfate reducer consortium. *Global Biogeochem. Cy.*, **8**(4), 451-463.

Holler, T., Widdel, F., Knittel, K., Amann, R., Kellermann, M.Y., Hinrichs, K.-U., Teske, A., Boetius, A., Wegener, G. 2011. Thermophilic anaerobic oxidation of methane by marine microbial consortia. *ISME J.*, **5**(12), 1946-1956.

Inagaki, F., Nunoura, T., Nakagawa, S., Teske, A., Lever, M., Lauer, A., Suzuki, M., Takai, K., Delwiche, M., Colwell, F.S. 2006. Biogeographical distribution and diversity of microbes in methane hydrate-bearing deep marine sediments on the Pacific Ocean Margin. *Proc. Natl. Acad. Sci. USA*, **103**(8), 2815-2820.

Joye, S.B., Boetius, A., Orcutt, B.N., Montoya, J.P., Schulz, H.N., Erickson, M.J., Lugo, S.K. 2004. The anaerobic oxidation of methane and sulfate reduction in sediments from Gulf of Mexico cold seeps. *Chemical Geology*, **205**(3), 219-238.

Kellermann, M.Y., Wegener, G., Elvert, M., Yoshinaga, M.Y., Lin, Y.-S., Holler, T., Mollar, X.P., Knittel, K., Hinrichs, K.-U. 2012. Autotrophy as a predominant mode of carbon fixation in anaerobic methane-oxidizing microbial communities. *Proc. Nat. Acad. Sci. USA,* **109**(47), 19321-19326.

Kimura, M. 1980. A simple method for estimating evolutionary rates of base substitutions through comparative studies of nucleotide sequences. *J. Mol. Evol.*, **16**(2), 111-120.

Knittel, K., Boetius, A. 2009. Anaerobic oxidation of methane: progress with an unknown process. *Annu. Rev. Microbiol.,* **63**(1), 311-334.

Knittel, K., Lösekann, T., Boetius, A., Kort, R., Amann, R. 2005. Diversity and distribution of methanotrophic archaea at cold seeps. *Appl. Environ. Microbiol.,* **71**(1), 467-479.

Krüger, M., Blumenberg, M., Kasten, S., Wieland, A., Känel, L., Klock, J.-H., Michaelis, W., Seifert, R. 2008. A novel, multi-layered methanotrophic microbial mat system growing on the sediment of the Black Sea. *Environ. Microbiol.*, **10**(8), 1934-1947.

Labrenz, M., Jost, G., Jürgens, K. 2007. Distribution of abundant prokaryotic organisms in the water column of the central Baltic Sea with an oxic-anoxic interface. *Aquat. Microb. Ecol.*, **46**(2), 177.

Lazar, C.S., Baker, B.J., Seitz, K., Hyde, A.S., Dick, G.J., Hinrichs, K.U., Teske, A.P. 2016. Genomic evidence for distinct carbon substrate preferences and ecological niches of Bathyarchaeota in estuarine sediments. *Environ. Microbiol.*, **18**(4), 1200-1211.

Lee, J.-W., Kwon, K.K., Azizi, A., Oh, H.-M., Kim, W., Bahk, J.-J., Lee, D.-H., Lee, J.-H. 2013. Microbial community structures of methane hydrate-bearing sediments in the Ulleung Basin, East Sea of Korea. *Mar. Petrol. Geol.*, **47**, 136-146.

Lloyd, K.G., Schreiber, L., Petersen, D.G., Kjeldsen, K.U., Lever, M.A., Steen, A.D., Stepanauskas, R., Richter, M., Kleindienst, S., Lenk, S. 2013. Predominant archaea in marine sediments degrade detrital proteins. *Nature*, **496**(7444), 215-218.

Losekann, T., Knittel, K., Nadalig, T., Fuchs, B., Niemann, H., Boetius, A., Amann, R. 2007. Diversity and abundance of aerobic and anaerobic methane oxidizers at the Haakon Mosby mud volcano, Barents Sea. *Appl. Environ. Microbiol.*, **73**(10), 3348-3362.

Marlow, J.J., Steele, J.A., Ziebis, W., Thurber, A.R., Levin, L.A., Orphan, V.J. 2014b. Carbonate-hosted methanotrophy represents an unrecognized methane sink in the deep sea. *Nat. Commun.*, **5**(5094), 1-12.

McGlynn, S.E., Chadwick, G.L., Kempes, C.P., Orphan, V.J. 2015. Single cell activity reveals direct electron transfer in methanotrophic consortia. *Nature*, **526**(7574), 531-535.

Meng, J., Xu, J., Qin, D., He, Y., Xiao, X., Wang, F. 2014. Genetic and functional properties of uncultivated MCG archaea assessed by metagenome and gene expression analyses. *ISME J.*, **8**, 650-659.

Meulepas, R.J.W., Stams, A.J.M., Lens, P.N.L. 2010. Biotechnological aspects of sulfate reduction with methane as electron donor. *Rev. Environ. Sci. Biotechnol.*, **9**(1), 59-78.

Meulepas, R.J.W., Jagersma, C.G., Khadem, A.F., Buisman, C.J.N., Stams, A.J.M., Lens, P.N.L. 2009. Effect of environmental conditions on sulfate reduction with methane as electron donor by an Eckernförde Bay enrichment. *Environ. Sci. Technol.*, **43**(17), 6553-6559.

Michaelis, W., Seifert, R., Nauhaus, K., Treude, T., Thiel, V., Blumenberg, M., Knittel, K., Gieseke, A., Peterknecht, K., Pape, T., Boetius, A., Amann, R., Jørgensen, B.B., Widdel, F., Peckmann, J., Pimenov, N.V., Gulin, M.B. 2002. Microbial reefs in the Black Sea fueled by anaerobic oxidation of methane. *Science*, **297**(5583), 1013-1015.

Milucka, J., Ferdelman, T.G., Polerecky, L., Franzke, D., Wegener, G., Schmid, M., Lieberwirth, I., Wagner, M., Widdel, F., Kuypers, M.M.M. 2012. Zero-valent sulphur is a key intermediate in marine methane oxidation. *Nature*, **491**(7425), 541-546.

Muyzer, G., Stams, A.J. 2008. The ecology and biotechnology of sulphate-reducing bacteria. *Nat. Rev. Microbiol.*, **6**(6), 441-454.

Niemann, H., Duarte, J., Hensen, C., Omoregie, E., Magalhaes, V.H., Elvert, M., Pinheiro, L.M., Kopf, A., Boetius, A. 2006a. Microbial methane turnover at mud volcanoes of the Gulf of Cadiz. *Geochim. Cosmochim. Ac.*, **70**(21), 5336-5355.

Niemann, H., Losekann, T., de Beer, D., Elvert, M., Nadalig, T., Knittel, K., Amann, R., Sauter, E.J., Schluter, M., Klages, M., Foucher, J.P., Boetius, A. 2006b. Novel microbial communities of the Haakon Mosby mud volcano and their role as a methane sink. *Nature*, **443**, 854-858.

Oni, O.E., Miyatake, T., Kasten, S., Richter-Heitmann, T., Fischer, D., Wagenknecht, L., Ksenofontov, V., Kulkarni, A., Blumers, M., Shylin, S. 2015. Distinct microbial populations are tightly linked to the profile of dissolved iron in the methanic sediments of the Helgoland mud area, North Sea. *Front. Microbiol.*, **6**(365), 1-15.

Orcutt, B., Boetius, A., Elvert, M., Samarkin, V., Joye, S.B. 2005. Molecular biogeochemistry of sulfate reduction, methanogenesis and the anaerobic oxidation of methane at Gulf of Mexico cold seeps. *Geochim. Cosmochim. Ac.*, **69**(17), 4267-4281.

Orphan, V.J., Hinrichs, K.U., Ussler, W., Paull, C.K., Taylor, L.T., Sylva, S.P., Hayes, J.M., Delong, E.F. 2001. Comparative analysis of methane-oxidizing archaea and sulfate-reducing bacteria in anoxic marine sediments. *Appl. Environ. Microbiol.*, **67**(4), 1922-1934.

Orphan, V.J., Ussler, W., Naehr, T.H., House, C.H., Hinrichs, K.-U., Paull, C.K. 2004. Geological, geochemical, and microbiological heterogeneity of the seafloor around methane vents in the Eel River Basin, offshore California. *Chem. Geol.*, **205**(3-4), 265-289.

Pachiadaki, M., G. , Lykousis, V., Stefanou, E., G., Kormas, K., A. 2010. Prokaryotic community structure and diversity in the sediments of an active submarine mud volcano (Kazan mud volcano, East Mediterranean Sea). *FEMS Microbiol. Ecol.*, **72**(3), 429-444.

Parkes, R.J., Cragg, B.A., Banning, N., Brock, F., Webster, G., Fry, J.C., Hornibrook, E., Pancost, R.D., Kelly, S., Knab, N., Jørgensen, B.B., Rinna, J., Weightman, A.J. 2007. Biogeochemistry and biodiversity of methane cycling in subsurface marine sediments (Skagerrak, Denmark). *Environ. Microbiol.*, **9**(5), 1146-1161.

Pruesse, E., Quast, C., Knittel, K., Fuchs, B.M., Ludwig, W., Peplies, J., Glöckner, F.O. 2007. SILVA: a comprehensive online resource for quality checked and aligned ribosomal RNA sequence data compatible with ARB. *Nucleic Acids Res.*, **35**(21), 7188-7196.

Qin, W., Amin, S.A., Martens-Habbena, W., Walker, C.B., Urakawa, H., Devol, A.H., Ingalls, A.E., Moffett, J.W., Armbrust, E.V., Stahl, D.A. 2014. Marine ammonia-oxidizing archaeal isolates display obligate mixotrophy and wide ecotypic variation. *Proc. Natl. Acad. Sci. USA*, **111**(34), 12504-12509.

Reeburgh, W.S. 2007. Oceanic methane biogeochemistry. *Chem. Rev.*, **107**(2), 486-513.

Ruff, S.E., Kuhfuss, H., Wegener, G., Lott, C., Ramette, A., Wiedling, J., Knittel, K., Weber, M. 2016. Methane seep in shallow-water permeable sediment harbors high diversity of anaerobic methanotrophic communities, Elba, Italy. *Front. Microbiol.*, **7**(374), 1-20.

Scheller, S., Yu, H., Chadwick, G.L., McGlynn, S.E., Orphan, V.J. 2016. Artificial electron acceptors decouple archaeal methane oxidation from sulfate reduction. *Science*, **351**(6274), 703-707.

Schloss, P.D., Westcott, S.L. 2011. Assessing and improving methods used in operational taxonomic unit-based approaches for 16S rRNA gene sequence analysis. *Appl. Environ. Microbiol.*, **77**(10), 3219-3226.

Seitaj, D., Schauer, R., Sulu-Gambari, F., Hidalgo-Martinez, S., Malkin, S.Y., Burdorf, L.D., Slomp, C.P., Meysman, F.J. 2015. Cable bacteria generate a firewall against euxinia in seasonally hypoxic basins. *Proc. Natl. Acad. Sci. USA*, **112**(43), 13278-13283.

Sipma, J., Meulepas, R.J.W., Parshina, S.N., Stams, A.J.M., Lettinga, G., Lens, P.N.L. 2004. Effect of carbon monoxide, hydrogen and sulfate on thermophilic (55°C) hydrogenogenic carbon monoxide conversion in two anaerobic bioreactor sludges. *Appl. Microbiol. Biotechnol.*, **64**(3), 421-428.

Sivan, O., Schrag, D.P., Murray, R.W. 2007. Rates of methanogenesis and methanotrophy in deep-sea sediments. *Geobiology*, **5**(2), 141-151.

Snaidr, J., Amann, R., Huber, I., Ludwig, W., Schleifer, K.H. 1997. Phylogenetic analysis and in situ identification of bacteria in activated sludge. *Appl. Environ. Microbiol.*, **63**(7), 2884-2896.

Sneath, P.H.A. 2005. Numerical Taxonomy. in: Brenner, D.J., Krieg, N.R., Staley, J.T., Garrity, G.M. (Eds.), *Bergey's Manual® of Systematic Bacteriology*. Springer US, USA, pp. 39-42.

Song, Z.-Q., Wang, F.-P., Zhi, X.-Y., Chen, J.-Q., Zhou, E.-M., Liang, F., Xiao, X., Tang, S.-K., Jiang, H.-C., Zhang, C.L., Dong, H., Li, W.-J. 2013. Bacterial and archaeal diversities in Yunnan and Tibetan hot springs, China. *Environ. Microbiol.*, **15**(4), 1160-1175.

Sulu-Gambari, F., Seitaj, D., Meysman, F.J.R., Schauer, R., Polerecky, L., Slomp, C.P. 2015. Cable bacteria control iron-phosphorus dynamics in sediments of a coastal hypoxic basin. *Environ. Sci. Technol.*, **50**(3), 1227-1233.

Tamura, K., Stecher, G., Peterson, D., Filipski, A., Kumar, S. 2013. MEGA6: Molecular Evolutionary Genetics Analysis Version 6.0. *Mol. Biol. Evol.*, **30**(12), 2725-2729.

Tavormina, P.L., Ussler, W., Joye, S.B., Harrison, B.K., Orphan, V.J. 2010. Distributions of putative aerobic methanotrophs in diverse pelagic marine environments. *ISME J.*, **4**(5), 700-710.

Thornburg, C.C., Zabriskie, T.M., McPhail, K.L. 2010. Deep-sea hydrothermal vents: potential hot spots for natural products discovery? *J. Nat. Prod.*, **73**(3), 489-499.

Thurber, A., Sweetman, A., Narayanaswamy, B., Jones, D., Ingels, J., Hansman, R. 2014. Ecosystem function and services provided by the deep sea. *Biogeosciences*, **11**(14), 3941-3963.

Timmers, P.H., Gieteling, J., Widjaja-Greefkes, H.A., Plugge, C.M., Stams, A.J., Lens, P.N.L., Meulepas, R.J. 2015. Growth of anaerobic methane-oxidizing archaea and sulfate-reducing bacteria in a high-pressure membrane capsule bioreactor. *Appl. Environ. Microbiol.*, **81**(4), 1286-1296.

Treude, T., Knittel, K., Blumenberg, M., Seifert, R., Boetius, A. 2005a. Subsurface microbial methanotrophic mats in the Black Sea. *Appl. Environ. Microbiol.*, **71**(10), 6375-6378.

Treude, T., Krüger, M., Boetius, A., Jørgensen, B.B. 2005b. Environmental control on anaerobic oxidation of methane in the gassy sediments of Eckernförde Bay (German Baltic). *Limnol. Oceanogr.*, **50**(6), 1771-1786.

Treude, T., Orphan, V.J., Knittel, K., Gieseke, A., House, C.H., Boetius, A. 2007. Consumption of methane and CO_2 by methanotrophic microbial mats from gas seeps of the anoxic Black Sea. *Appl. Environ. Microbiol.*, **73**(7), 2271-2283.

Vasquez-Cardenas, D., van de Vossenberg, J., Polerecky, L., Malkin, S.Y., Schauer, R., Hidalgo-Martinez, S., Confurius, V., Middelburg, J.J., Meysman, F.J.R., Boschker, H.T.S. 2015. Microbial carbon metabolism associated with electrogenic sulphur oxidation in coastal sediments. *ISME J.*, **9**(9), 1966-1978.

Vigneron, A., Cruaud, P., Pignet, P., Caprais, J.-C., Cambon-Bonavita, M.-A., Godfroy, A., Toffin, L. 2013a. Archaeal and anaerobic methane oxidizer communities in the Sonora Margin cold seeps, Guaymas Basin (Gulf of California). *ISME J.*, **7**(8), 1595-1608.

Vigneron, A., Cruaud, P., Pignet, P., Caprais, J.-C., Gayet, N., Cambon-Bonavita, M.-A., Godfroy, A., Toffin, L. 2013b. Bacterial communities and syntrophic associations involved in AOM process of the Sonora Margin cold seeps, Guaymas basin. *Environ. Microbiol.*, **16**(9), 2777-2790.

Wagner, M. 2015. Microbiology: Conductive consortia. *Nature*, **526**(7574), 513-514.

Wagner, M., Amann, R., Lemmer, H., Schleifer, K.-H. 1993. Probing activated sludge with oligonucleotides specific for proteobacteria: inadequacy of culture-dependent methods for describing microbial community structure. *Appl. Environ. Microbiol.*, **59**(5), 1520-1525.

Wang, F.-P., Zhang, Y., Chen, Y., He, Y., Qi, J., Hinrichs, K.-U., Zhang, X.-X., Xiao, X., Boon, N. 2014. Methanotrophic archaea possessing diverging methane-oxidizing and electron-transporting pathways. *ISME J.*, **8**(5), 1069-1078.

Wankel, S.D., Adams, M.M., Johnston, D.T., Hansel, C.M., Joye, S.B., Girguis, P.R. 2012. Anaerobic methane oxidation in metalliferous hydrothermal sediments: influence on carbon flux and decoupling from sulfate reduction. *Environ. Microbiol.*, **14**(10), 2726-2740.

Widdel, F., Bak, F. 1992. Gram negative mesophilic sulfate reducing bacteria. in: Balows, A., Truper, H., Dworkin, M., Harder, W., Schleifer, K.H. (Eds.), *The prokaryotes: a handbook on the biology of bacteria: ecophysiology, isolation, identification, applications*,Vol. II. Springer New York, USA, pp. 3352-3378.

Zhang, Y. 2011. Microbial processes in a simulated cold seep ecosystem. PhD thesis, Ghent University. Belgium, pp. 61-84.

Zhang, Y., Arends, J.B., Van de Wiele, T., Boon, N. 2011. Bioreactor technology in marine microbiology: from design to future application. *Biotech. Adv.*, **29**(3), 312-321.

Zhang, Y., Henriet, J.-P., Bursens, J., Boon, N. 2010. Stimulation of in vitro anaerobic oxidation of methane rate in a continuous high-pressure bioreactor. *Bioresource. Technol.*, **101**(9), 3132-3138.

CHAPTER 5

Enrichment of Anaerobic Methane Oxidizing ANME-1 from Ginsburg Mud Volcano (Gulf of Cadiz) Sediment in a Biotrickling Filter

The modified version of this chapter is published as:

Bhattarai S., Cassarini C., Rene E.R ., Zhang Y., Esposito G.and Lens P.N.L. (2018) Enrichment of sulfate reducing anaerobic methane oxidizing community dominated by ANME-1 from Ginsburg Mud Volcano (Gulf of Cadiz) sediment in a biotrickling filter. *Bioresource Technology, 259* (2018), 433-441

Abstract

This study aimed to enrich anaerobic methane oxidizers (ANME) present in sediment from the Ginsburg Mud Volcano (Gulf of Cadiz) in a polyurethane foam packed biotrickling filter (BTF). The inoculated sediment was stored for 4 years at 4°C under methane atmosphere prior to inoculation in BTF. The BTF was operated in continuous supply of methane and/or nitrogen for 380 days. Under the methane-rich environment in the BTF, methane and sulfate consumption were coupled to total dissolved sulfide production at approximately equimolar ratios. Sulfate reduction with simultaneous sulfide production accumulating up to 7 mM after 200 days, indicates the development of anaerobic oxidation of methane (AOM) coupled to sulfate reduction. Illumina high-throughput sequence analysis of 16S rRNA genes showed that after 248 days of bioreactor operation, nearly 50% of the archaeal sequences belonged to the anaerobic methanotrophs (ANME) clades, including ANME-1 (42%) and ANME-2 (8%). Other major archaeal clades in the bioreactor enrichment were the Miscellaneous Crenarchaeotic Group (33%). A relatively shorter BTF start-up period in comparison to other AOM enriching bioreactors and high rate of AOM induced sulfate reduction were achieved in this study. The proliferation of ANME was supported by the packing material of the BTF.

5.1　Introduction

The anaerobic oxidation of methane (AOM) coupled to the reduction of sulfate (AOM-SR, Eq. 5.1) is a microbial process that largely prevents the emission of methane to the atmosphere in marine environments (Knittel & Boetius, 2009; Reeburgh, 2007).

$$CH_4 + SO_4^{2-} \rightarrow HCO_3^- + HS^- + H_2O \quad \Delta G° = \text{-17 kJ mol}^{-1} CH_4 \quad \text{Eq. 5.1}$$

AOM-associated microorganisms have been extensively studied using biogeochemical and microbiological approaches. The euryarchaeal anaerobic methanotrophs, also known as the anaerobic methane oxidizers (ANME), perform AOM either solely or in syntrophic association with deltaproteobacterial sulfate-reducing bacteria (SRB) (Knittel & Boetius, 2009). ANME are phylogenetically related to various groups of methanogenic *Archaea* and grouped into three distinct clades, i.e., ANME-1, ANME-2 and ANME-3, respectively (Boetius et al., 2000; Hinrichs et al., 1999; Knittel et al., 2005; Niemann et al., 2006a; Niemann et al., 2006b). ANME-2 and ANME-3 are clustered within the order *Methanosarcinales*, while ANME-1 belongs to a new order which is distantly related to the orders *Methanosarcinales* and *Methanomicrobiales*. ANME-1 has been retrieved from a wide variety of marine habitats having temperature variations between 4°C and 70°C (Holler et al., 2011; Vigneron et al., 2013a), whereas ANME-2 and ANME-3 are mostly found in environments with temperatures < 20°C (Hinrichs et al., 2000; Losekann et al., 2007; Marlow et al., 2014; Niemann et al., 2006b; Oni et al., 2015; Orphan et al., 2001; Ruff et al., 2016).

The low solubility of methane, maintenance of anaerobic conditions and slow growth rates of ANME are major constraints for the enrichment of ANME. Besides, ANME grown in the laboratory under simulated conditions represent only a small fraction of the total diversity that exists in nature. Several attempts have been made to achieve ANME enrichments and higher AOM activities in batch incubations (Holler et al., 2011; Nauhaus et al., 2007; Schreiber et al., 2010), membrane bioreactors (Meulepas et al., 2009a) and continuously operated bioreactors (Aoki et al., 2014; Girguis et al., 2005; Wegener & Boetius, 2008). Provision of continuous methane supply and biomass retention can enhance AOM activity by 20-200 fold of the activity obtained in the batch systems (Meulepas et al., 2009a; Nauhaus et al., 2007). As reported in the literature, continuously operated bioreactors maintained at standard atmospheric conditions usually require longer start-up periods and exhibit lower AOM activities compared to high pressure bioreactors (Deusner et al., 2009; Timmers et al., 2015; Zhang et al., 2011) or enrichment under *in situ* conditions (Jagersma et al., 2009; Timmers et al., 2015). Nevertheless, the high operational and maintenance costs of high pressure bioreactors and their elaborate design and safety requirements can be a hindrance for their application.

In this study, a newly designed biotrickling filter (BTF) with continuous methane supply was tested for the enrichment of ANME and associated microbial community using deep sea methane-seep sediment from Ginsburg mud volcano of the Gulf of Cadiz (the Atlantic Ocean) as inoculum. The key design and operational factors of the BTF were considered, such as biomass retention, gas-solid mass transfer and reduction of sulfide toxicity. Considering the low solubility of methane, it is preferred to have packing materials with high porosity in order to enhance the gas mass transfer by retaining the gas in pores (Avalos Ramirez et al., 2012; Estrada et al., 2014a; Estrada et al., 2014b). Polyurethane foam can retain the biomass in the highly connected pore spaces and effectively trap as well as release the sorbed methane and other gaseous end products (Aoki et al., 2014). Therefore, this study investigated the AOM activity of long term incubated Ginsburg mud volcano sediment in a BTF reactor. Further, the change in microbial community was tracked by Illumina Miseq method during BTF operation. The obtained results of AOM activity along with the microbial analysis were compared with other bioreactor based AOM studies.

5.2 Materials and methods

5.2.1 Source of biomass

Sediment samples were obtained from the Ginsburg mud volcano, Gulf of Cadiz (Spain), from 53 cm to 153 cm below the ocean floor. The coring was performed in 2009 by the MicroSYSTEMS cruise of R/V Pelagia from the crater of the mud volcano (35° 22.431'N; 07° 05.291'W, water depth ca. 910 m) (Zhang et al., 2010). For the sampling, a plastic liner was inserted into a Box Core (40 cm deep) that contained pieces of gas hydrate (0.5 to 5 cm in length). The core was immediately capped and stored at 4°C in trilaminate polyetherimide coated aluminum bags (KENOSHA C.V., Amstelveen, The Netherlands) under nitrogen atmosphere (Zhang et al., 2010). The original sediment was diluted 2 times with artificial seawater and homogenized wet sediment mixture was prepared. Prior to the inoculation in the BTF, the sediment was stored at 4°C under methane atmosphere for four years. The redox condition was maintained in the sediment during the storage by regular monitoring of added color indicator (resazurin) and regular replacement of headspace methane.

The Gulf of Cadiz is located in the eastern Atlantic ocean, northwest of the Strait of Gibraltar, along the Spanish and Portuguese continental margins (Niemann et al., 2006b). The area is well known for several mud volcanoes provinces and the Ginsburg mud molcano is located within the Western Moroccan mud volcano field of the Gulf of Cadiz. These mud volcanoes are known for their environmental heterogeneity which is directly related to the cold seep environments with an AOM rate of 10-104 nmol cm^{-2} d^{-1} and sulfate reduction rate of 158-189 nmol cm^{-2} d^{-1}

(Niemann et al., 2006b), which is in the low to mid range compared to other known AOM sites such as the Black sea carbonate chimney, Haakon Mosby mud volcano, Eel river basin and Gulf of Mexico hydrate., The *in situ* or *ex situ* AOM-SR rate for Ginsburg mud volcano has however not yet been estimated.

5.2.2 Mineral medium composition

The artificial seawater medium was prepared as reported in a previous study (Zhang et al., 2010), and contained (per L of demineralised water): NaCl (26 g), KCl (0.5 g) $MgCl_2 \cdot 6H_2O$ (5 g), NH_4Cl (0.3 g), $CaCl_2 \cdot 2H_2O$ (1.4 g), Na_2SO_4 (1.43 g for 10 mM of sulfate), KH_2PO_4 (0.1 g), trace element solution (1 ml), 1 M $NaHCO_3$ (30 ml), vitamin solution (1 ml), thiamin solution (1 ml), vitamin B_{12} solution (1 ml), 0.5 g L^{-1} resazurin solution as a redox indicator (1 ml) and 0.5 M Na_2S solution (1 ml). The pH was adjusted to 7.0 with sterile 1 M Na_2CO_3 or H_2SO_4 solution.

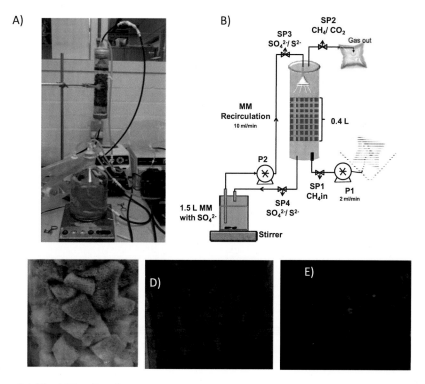

Figure 5.1 Biotrickling filter for the enrichment of ANME: A) BTF schematic representation. and B) BTF setup in the laboratory and polyurethane foams inside the BTF column during different operation periods, viz., C) at 50 days D) at 120 days and E) at 248 day. Here, MM refers to artificial seawater medium, P refers to pump and SP refers to the sampling port with the respective numbers.

5.2.3 BTF set-up

The BTF (Figure 5.1) was mounted in a cylindrical polyvinyl chloride (PVC) pipe, (height: 32 cm, diameter: 56 mm). The packed volume of the reactor was 0.4 L, filled with polyurethane foam cubes of 1 cm × 1 cm × 1 cm, 98% porosity and a density of 28 kg m^{-3}. A circular acrylic sieve plate (pore size of 3.5 mm) was used to hold the polyurethane foams in the BTF. Another circular sieve of 4.0 mm pore size was placed at the top of the BTF to distribute the liquid medium uniformly over the entire cross sectional area of the packing material (Figures 5.1A and 5.1B). The BTF was sealed air-tight to prevent leakage or air intrusion during its operation.

5.2.4 Experimental design

The BTF was inoculated by 120 ml of homogenized wet sediment mixture. The operational temperature of the BTF was 20 ($\pm$2)°C, which was slightly higher than the average *in situ* temperature (15°C) of the Gulf of Cadiz. The pH and salinity (7.5 and 30 ‰) were maintained at values that were comparable with the sampling site of Ginsburg mud volcano (Niemann et al., 2006b). The BTF was operated in sequential fed-batch mode of the influent, but with continuous recirculation for liquid phase. Methane gas (100% methane) stored in Tedlar bags was continuously supplied to the bioreactor using a peristaltic pump at a gas flow rate of about 2 ml min^{-1} and vented out through an exit port. The estimated empty bed residence time of methane in the BTF was 200 min.

The mineral medium (with 10-18 mM sulfate) was recirculated in the BTF and passed from the top of the bioreactor at a flow rate of 10 ml min^{-1}. During the entire operational period of the BTF, the redox was maintained negative and anaerobic as the rezasurin containing mineral medium remained colorless. The BTF operation was divided in six phases as shown in Table 1. The first 120 days of BTF operation corresponded to the reactor start-up phase, i.e., when the microbial community was in the acclimatization stage (phase I). Phase II (from day 121 to day 248) was the methane supply phase or ANME enrichment phase. To check the AOM activity in the absence of excess CH$_4$ supply, the operational mode of the BTF was changed by supplying N$_2$ instead of CH$_4$ intermittently from days 248 to 269 (phase III) and again from days 289 to 338 (phase V). CH$_4$ supply was retained in between the two N$_2$

supply phases (phase III and phase V) from days 270 to 289 (phase IV) and days 339 to 380 (phase VI) of reactor operation.

Mineral medium was refreshed in the BTF in all phases of BTF operation. Mineral medium was refreshed in the reactor in order to avoid potential sulfide toxicity for six times after phase I of BTF operation as indicated by black arrows in Figure 5.2. In the first phase of BTF operation, the mineral medium (with almost 18 mM of sulfate) was changed to make sure there was no oxygen intrusion or bacterial contamination in the mineral medium. In the later phases (phases II to VI, refer Table 1), the major criterium to determine the change of mineral medium was the total dissolved sulfide concentration, i.e. when the total dissolved sulfide concentration was ~ 5 mM, the mineral medium in the holding tank and BTF was changed to prevent potential sulfide toxicity to ANME. Toxic effects of total dissolved sulfide have been observed beyond a concentration of 4 mM for high pressure incubations with Captain Arutyunov Mud Volcano (Zhang et al., 2011) and 2.5 mM of sulfide has already shown to be toxic in a continuously operated membrane bioreactor used to enrich the ANME from Eckernförde Bay (Meulepas et al., 2009b). Liquid sample (~ 5ml) for sulfate, sulfide and pH measurement was extracted from sampling port (SP4, Figure 5.1) while gas sample was extracted from both SP1(gas in-let) and SP2 (gas out-let).

5.2.5 Microbial analysis

The microbial community for this study was analyzed by Illumina Miseq high throughput sequencing method (Caporaso et al., 2012). The sediment slurry inoculated in BTF and the enriched biomass from the active phase at 248 days were stored in -20°C for microbial community analysis.

5.2.6 DNA extraction

DNA was extracted by using a FastDNA® SPIN Kit for soil (MP Biomedicals, Solon, OH, USA) by following the manufacturer's protocol. Almost 0.5 g of the sediment was used for DNA extraction from the initial inoculum and almost 0.5 ml of liquid obtained by washing the polyurethane foam with nuclease free water was used for DNA extraction of enriched slurry. The extracted DNA was quantified by a Nanodrop ND-1000 Spectrophotometer (Thermo Fisher Scientific, MA, USA) and the quality of DNA was checked by gel electrophoresis with 1% agarose.

5.2.7 PCR amplification for 16S rRNA genes

Archaeal and bacterial 16S rRNA genes in variable region 4 (V4) were amplified for the inoculum and enriched biomass. For the enriched biomass, universal archaeal primer pair with identification tags Uni516F (Takai & Horikoshi, 2000) and arch855R (Gray et al., 2002) were used. However, for the case of the initial inoculum, several strategies were applied for the PCR of *Archaea*, as the amplification was not achieved with the above mentioned primer pair. In this case, the Archaeal genes from the initial inoculum were amplified by nested PCR, in which the DNA was first amplified with universal archaeal primer pair Arc21F and Arc958R as described by DeLong (1992). Then, amplicon was used as template for V4 amplification by primer pair UNI519F and Arc806R (Song et al., 2013) as nested PCR. For both the inoculum and the enriched biomass, bacterial genes were amplified by using the universal primer pairs Bac533F and Bac802R, respectively (Song et al., 2013).

The PCR reaction mixture (50 µl) contained 2 µl of DNA template (~ 100 ng), 1 × PCR buffer with 1.5 mM MgCl$_2$, 0.2 mM of deoxynucleoside triphosphates (dNTPs), 2.5 mg L^{-1} of Bovine Serum Albumin (BSA), 0.4 mM of each primer, 1.25 U of Ex Taq DNA polymerase (all the chemicals were purchased from Takara products, Japan) and nuclease free water. An initial denaturation step of 5 min at 94°C was followed by 30 cycles at 94°C for 40 sec, 55°C for 55 sec and 72°C for 40 sec. The final extension step was 72°C for 10 min. However, 20 cycles were used for the first step of nested PCR and the annealing temperature for the primer pairs UNI516F and arch855R was maintained at 60°C. PCR amplicons were purified using the E.Z.N.A.® Gel Extraction Kit by following the manufacturer's protocol (Omega Biotek, USA).

5.2.8 Illumina Miseq data processing

The purified DNA amplicons were sequenced by an Illumina HiSeq 2000 (Illumina, San Diego, CA, USA) and analyzed as detailed previously (He et al. 2016). A total of 40000 (± 20000) sequences were assigned to *Archaea* and *Bacteria* each by examining the tags assigned to the amplicons. Chimera elimination, sequence analysis and classification were performed by mothur project bioinformatics tools (Schloss & Westcott, 2011), as described by He et al. (2016). Taxonomic classification of the retrieved sequences was performed by using the SILVA database (Pruesse et al., 2007).

5.2.9 Cells fixation and visualization

The enriched slurry samples were withdrawn from the reactor at the end of the study period (380 days). Cells were fixed in 1% paraformaldehyde (PFA) and phosphate buffer saline (PBS) overnight at 4°C. After fixation, cells were washed three times in PBS and stored in 50% ethanol/PBS (v/v). Samples were filtered through 0.22 μm pore sized polycarbonate filters (GTTP type, Millipore), supported by cellulose acetate filters (0.45 μm pore size, HAWP type, Millipore), the filters were stored at - 20°C.

Cells were visualized and identified by catalyzed reporter deposition-fluorescence in situ hybridization (CARD-FISH) and fluorescence in situ hybridisation FISH. FISH from the BTF enrichment samples might underestimate microbial interactions as the double hybridization of ANME and SRB was not performed. Nevertheless, the main objective of FISH was to check the ANME-1 cells in the enrichment biomass. About 0.5 ml of enriched biomass was obtained by washing polyurethane foam with nuclease free water and FISH was performed by following standard protocol described previously (Bhattarai et al., 2017). The biomass was hybridized with the Arc-FAM probe (Stahl and Amann, 1991), a mixture of Cy3-labeled ANME probes: ANME-1 350 (Boetius et al., 2000), ANME-2 538 (Treude et al., 2005), ANME-3 1249 (Niemann et al., 2006b) and Cy5-labelled *DBB* 660 (Niemann et al., 2006b). Following the removal of unbound probes, cells were counterstained with 4',6-diamidino-2-phenylindole (DAPI) (Wagner et al., 1993). The hybridization and microscopic visualization of fluorescent cells was performed as described previously (Snaidr et al., 1997).

Prior to CARD-FISH analysis, filters were dipped in low-gelling-point agarose (0.1% w/v in Milli-Q water), dried on a petri dish at 37°C and dehydrated in 96% (v/v) ethanol. For bacterial cell wall permeabilization, samples were treated with 10 mg ml⁻^{1} of lysozyme, 50 mM EDTA and 0.1 M Tris-HCl), at 37°C for 1 h, washed with MilliQ water and subsequently treated with achromopeptidase (60 U ml^{-1} in 0.01 M NaCl and 0.01 M Tris-HCl, pH 8). Filters were washed 3 times in MilliQ water. For archaeal cell wall permeabilization, filters were incubated with sodium dodecyl sulfate (SDS) and proteinase K solution as described by Holler et al. (2011). For all the samples, endogenous peroxidases were inactivated by incubation in 0.1% H_2O_2 for 1-2 min at room temperature and then the filters were washed twice in MilliQ water and dehydrated with 50, 80 and 96% (v/v) ethanol (1 min each).

CARD-FISH with horseradish peroxidase (HRP)-labeled oligonucleotide probes and tyramide signal amplification was done according to the protocol described elsewhere (Pernthaler et al., 2002; Pernthaler et al., 2004). The microorganisms were visualized using archaeal and bacterial HRP-labeled oligonucleotide probes ARCH915 (Stahl and Amann, 1991) and EUB338-I-III (Daims et al., 1999), respectively, at a formamide concentration of 35% (v/v hybridization buffer). The probes DSS658 (Boetius et al., 2000) and ANME-2 538 (Schreiber et al., 2010) were used for the detection of Desulfosarcina-Desulfococcus sp. and ANME-2 Archaea, respectively, using a formamide concentration of 50%. Oligonucleotide probes were purchased from Biomers (Germany).

For dual CARD-FISH, peroxidases of initial hybridizations were inactivated as described previously by Holler et al. (2011). Tyramide amplification was performed using the fluorochromes Alexa Fluor 488 and Alexa Fluor 594, prepared as described by Pernthaler et al. (2004). Finally, all the cells were stained with 4', 6'-diamidino-2-phenylindole (DAPI) and analyzed using a epifluorescence microscope (Carl Zeiss, Germany).

5.2.10   Analytical method

pH was measured with a Metrohm pH meter (Metrohm Applikon B.V., Schiedam, the Netherlands) and a pH electrode (SenTix WTW, Amsterdam, the Netherlands). One volume of sample (0.5 ml) was diluted to one volume of 0.5 M sodium hydroxide to raise the pH in order to prevent the volatilization of total sulfide. Sulfate was analyzed using an Ion Chromatograph system (Dionex-ICS-1000 with AS-DV sampler) as described previously (Villa-Gomez et al., 2011). Total dissolved sulfide was analyzed using the methylene blue method (Siegel, 1965). The amount of total dissolved sulfide measured accounted for all the cumulative dissolved sulfide species (H_2S, HS^- and S^{2-}).

Methane and carbon dioxide concentrations in the gas inlet and outlet of BTF were measured by gas chromatography (GC 3800, VARIAN). The gas chromatograph was equipped with a PORABOND Q column (25 m × 0.53 mm × 10 μm) and a thermal conductivity detector. The following conditions were used in the gas chromatograph: carrier gas - helium (15 Psi), flow rate - 0.5 ml min^{-1}, oven temperature - 25°C and the gas injection volume - 0.5 ml. For each sampling, gas measurements were performed in duplicates and a standard deviation lower than 0.5% in the chromatograph peak area was used in the calculation. Standard gas mixtures of methane and carbon dioxide were measured every time along with sample measurements.

5.3 Results

5.3.1 AOM activity during BTF operation

During the start-up phase (phase I), the microbial community was in its acclimatization stage and the microbial AOM-SR activity was almost negligible. Sulfate and carbon dioxide concentrations remained almost stable, while methane consumption was below 0.5 mM (Table 5.1 and Figure 5.2). Total dissolved sulfide production was not detected in phase I. pH was around 7.0 during the initial phase of reactor operation and started to increase towards the end of phase I, increasing up to 8.3 in the later phases.

pH shifted to alkaline when microbial activity started in phase II, indicating the production of biocarbonate alkalinity as a result of the AOM-SR process (Figure 5.2A). During phase II, active AOM-SR was observed. Almost 10 mM of sulfate was reduced with simultaneous cumulative sulfide production of 8 mM (Figure 5.2B). The sulfate reduction profiles during phase II showed sulfate consumption coupled to sulfide production at approximately equimolar ratios, which is consistent with the stoichiometry of AOM-SR. Similarly, There was simultaneous increment of methane consumption with sulfate reduction in phase II with the average sulfate reduction to methane consumption ratio of almost 1:1.4 (Figure 5.2C). A maximum rate of methane consumption was observed during this phase with almost 3 mM of methane consumption at the rate of 1 mM d^{-1} in between day 160 and 200. However, carbon dioxide production was almost half or lower than the half of the methane consumption. The maximum amount of carbon dioxide production was observed on day 210 with carbon dioxide concentration of 1.5 mM, whereas the amount of methane consumption measured on that day was 2.6 mM (Figure 5.2 C).

During phase III, sulfate reduction and total dissolved sulfide production continued even when methane supply was stopped and nitrogen was supplied to the BTF. The sulfate reduction rate was 0.28 mM d^{-1} during this phase, which is almost similar to the sulfate reduction rate of phase II (Table 5.1). However with the supply of nitrogen instead of methane during phase III, the carbon dioxide production remained lower than 0.5 mM with oscillating trend.

Table 5.1 *Maximum volumetric rates of sulfate reduction, sulfide production, methane oxidation and carbon dioxide production during different phases of BTF operation*

Phases of reactor operation	Duration in days	Sulfate reduction	Sulfide production	Methane oxidation	Carbon dioxide production
		Maximum volumetric rate, mM d^{-1}			
Phase I, startup phase	0-120	0.0075	0.00021	0.0037	0.0037
Phase II, CH$_4$ supply	120-248	0.36	0.3	0.83	0.31
Phase III, N$_2$ supply	248-269	0.28	0.25	0.28	0.8
Phase IV, CH$_4$ supply	269-289	1.3	0.78	1.05	0.79
Phase V, N$_2$ supply	289-338	0.34	0.46	0.21	0.20
Phase VI, CH$_4$ supply	338-380	0.40	0.37	0.38	0.18

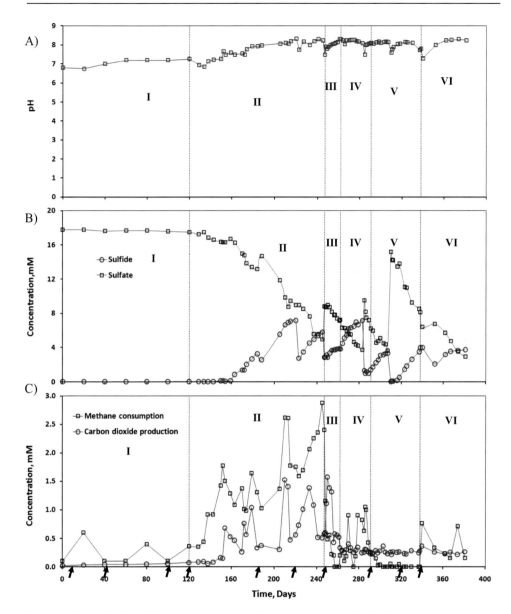

Figure 5.2 A) pH, B) Sulfate reduction and sulfide production and C) Methane concentration and carbon dioxide concentration changes during operation of BTF for the Ginsburg mud volcano enrichment of ANME at ambient temperature and pressure. The black arrows indicate the time at which the nutrient medium was changed.

The sulfate reduction as well as total dissolved sulfide production rate was the highest in phase IV with a sulfate reduction rate of 1.3 mM d^{-1}. In agreement with the sulfate reduction rate, the methane consumption rate was the highest in phase IV

(Table 5.1 and Figure 5.2). However, after the nitrogen supply test, total methane consumption and carbon dioxide production was reduced by four times compared to phase II. Carbon dioxide production remained almost stable with 0.2-0.3 mM at this phase (Figure 5.2C). Although the maximum methane consumption rate was achieved during phase IV (Table 5.1), the total amount of methane consumed remained below 1 mM in this phase (Figure 5.2C).

Similar to the phase III, the sulfate reduction and simultaneous total dissolved sulfide production was observed in phase V. However, the sulfate reduction rate was almost half compared to the values achieved in phase IV (Table 5.1). There was no methane consumption during phase V and also carbon dioxide production was stable with less than 0.3 mM of carbon dioxide production. Methane supply was further retained in phase VI, the sulfate reduction was observed as in other phases. However, methane consumption and simultaneous carbon dioxide production was not achieved as in phase II so nitrogen supply test might have influenced the AOM biomass.

5.3.2 Archaeal 16S rRNA relative abundance

A total of 40,000 ($\pm$ 10,000)sequences reads were obtained for the archaeal 16S rRNA genes amplicon from the sediment used as inoculum for the BTF via high-throughput sequencing. The analysis showed the identity up to the order level with the presence of major archaeal phylotypes, i.e., *Euryarchaeota* (95%), *Thaumarchaeota* (4%) and unclassified *Archaea* (1%). The majority of euryarcheotal sequences were belonging to the clade *Thermococci* with a relative abundance of 85%. The 16S rRNA genes abundance analysis showed that the sediment used as inoculum for the BTF reactor contained mainly *Thermococcales* (85%) and Marine Group II (10%) as archaeal components (Figure 5.3A), whereas none of the ANME types were detected using the method employed in this study. However, these archaeal communities detected in the initial inoculum were possibly biased by the nested PCR approach applied before sequencing.

After operation of BTF for 248 days, the majority of the archaeal sequences were affiliated with ANME clades. Similar to the case of the inoculum, the archaeal sequences from the enrichment also showed identity up to order level with the presence of major archaeal phylotypes, i.e., *Euryarchaeota* (55%), *Crenarchaeota*

(34%), *Thaumarchaeota* (9.8%) and *unclassified Archaea* (0.7%). *Euryarchaeota* and *Crenarcharchaeota* were the most abundant phyla in the BTF biomass.

The majority of the sequences from euryarchaeotic phyla were affiliated to ANME, which was also 50% of the total sequence analyzed (Figure 5.3A). Among the ANME clades, the most abundant clade was ANME-1, more specifically ANME-1b type (41%), followed by ANME-2b (8%) and ANME-2a (1.2%), whereas other ANME clades such as ANME-3 or ANME-2c and ANME-2d were not detected. *Methanomicrobia*, a class of *Euryarchaeota* related to methanogens and the associates of ANME-1 were also retrieved from the enrichment (Figure 5.3A). However, their relative abundance was less than 0.5% (Figure 5.3A). Furthermore, other clades of methanogens such as the *Methanosarinales* and *Meathanococcoides* order of *Euryarchaeota* were found, but with an abundance of only about 4% in total.

The second larger cluster of *Archaea* in enriched BTF biomass was the Miscellaneous Crenarchaeotic Group (MCG), also known as *Bathyarchaeaota* recently (Meng et al., 2014) Within the MCG, the majority of the sequences were related to uncultured *Crenarchaeota* with a relative abundance of 32%. Therefore, those uncultured MCG could not be distinguished into the classes and genus levels. Further, *Thaumarchaeota* phylum related *Archaea* were also present with less than 10% in abundance. The majority of *Thaumarchaeota* were affiliated with Marine Benthic Group B and Group C3 types of *Archaea,* which are often retrieved from marine sediments. Marine Group-I type *Thaumarcheaota* were also present in the reactor enrichment; however, their relative abundance was lower than 1%.

5.3.3 Bacterial 16S rRNA relative abundance

Bacterial 16S rRNA genes from the inoculum were distributed among different phyla, including *Bacteroidetes, Firmicutes, Planctomycetes, Proteobacteria* and *Spirochaetes*. The most abundant bacterial phyla were *Proteobacteria* with more than 86% sequences affiliated to it, followed by *Bacteroidetes* (5%) and *Firmicutes* (6%).

The *Proteobacteria* phylum was diversified with *Alphaproteobacteria, Betaproteobacteria, Gammaproteobacteria, Deltaproteobacteria* and *Epsilonproteobacteria* (Figure 5.3B). The relative abundance of *Deltaproteobacteria* phyla, which incorporates sulfate reducing bacteria, was about 1% of the total bacterial sequences retrieved, with the majority (~ 40%) of *Gammaproteobacteria*. The *Gammaproteobacteria* incorporating sulfide oxidizing microbial community of *Thiotrichales* was with the highest relative abundance in the inoculum, containing almost 45% of the total sequences.

Another major group of the *Proteobacteria* was the *Epsilonproteobacteria,* consisting mainly of the *Helicobacteraceae* family with a relative abundance of almost 20%. The *Rhodobacteraceae* family of *Alphaproteobacteria* was also retrieved with a 10% relative abundance (Figure 5.3B). Similar to the inoculum, the bacterial sequences from the enriched biomass (on day 248) also showed the presence of diverse bacterial phyla. The bacterial phyla with the highest relative abundance were *Proteobacteria* with more than 80% sequences affiliated to it, followed *by Bacteroidetes* (11%) and *Firmicutes* (6%).The *Proteobacteria* phylum was diversified with *Alphaproteobacteria,* *Betaproteobacteria,* *Gammaproteobacteria,* *Deltaproteobacteria and Epsilonproteobacteria* (Figure 5.3B). Among the *Proteobacteria,* the phylotypes with higher relative abundance were the *Alphaproteobacteria,* *Betaproteobacteria* and *Gammaproteobacteria.* *Gammaproteobacteria* were almost 40% in relative abundance consisting of mainly *Pseudomonadales,* whereas for the inoculum the most abundant order was the *Thiotrichales*.

The *Burkholderiales* order of the *Betaproteobacteria* was the major clade in the enriched biomass with a relative abundance of 30%, although the clade was less than 2% of the total bacterial sequences in the inoculum. Unlike other AOM enrichments, the enrichment in the BTF showed a very low relative abundance of *Deltaproteobacteria*, i.e., almost 1% which includes the SRB clades of *Desulfobacterales* and *Desulfuromundales* (Figure 5.3B).

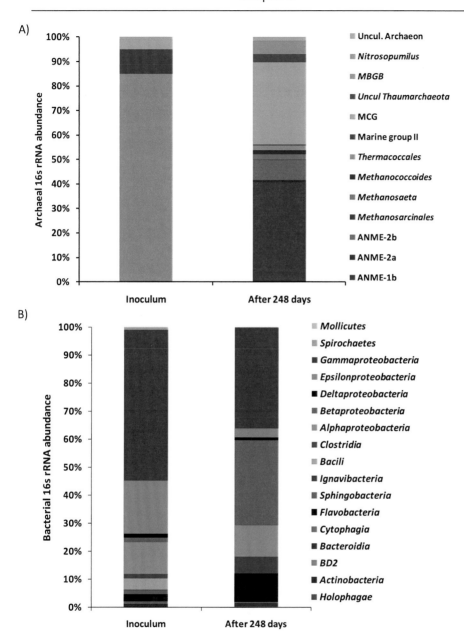

Figure 5.3 A) Archaeal and B) Bacterial community composition of Ginsburg mud volcano sediment after 248 days of BTF operation (end of phase II) as derived by Illumina Miseq sequencing of PCR amplicons.

5.3.4 Microbial visualization

Figure 5.4 shows the cells visualized by CARD-FISH using different permeabilization depending on the targeted cells. Figure 5.4A is a representative image for the abundance of archaeal versus bacterial cells, showing that after 380 days of reactor operation over 90% of the cells are bacterial cells.

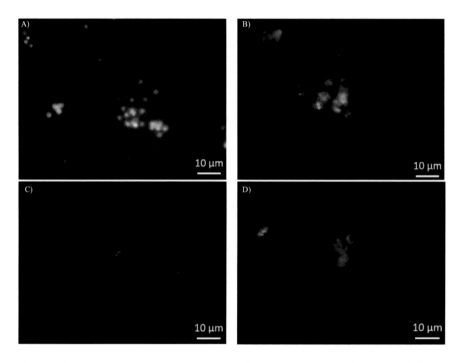

Figure 5.4 Visualization of cells at the end of BTF operation by CARD-FISH A) bacterial cells in orange and archaeal cells in violet hybridized by EUB338-I-III and ARCH915 probe, respectively, B) All cells stained by DAPI in blue and Desulfosarcinales in green hybridized by DSS658 probe, C) Archaeal cells in violet and ANME-2 cells in red hybridized by ARCH915 and ANME2-538, respectively, and D) Green cells hybridized by DSS658 and red cells hybridized by ANME2-538 probe.

In agreement with Illumina Miseq analysis, the *Desulfosarcinales* phylotypes were rarely present in the analysed BTF biomass (Figure 5.4B). ANME-2 cells were visualized by CARD-FISH and the results were in agreement with Illumina Miseq results and around 10% of the archaeal cells were identified as ANME-2 (Figure 5.4C). Double hybridization with ANME-2 and *Desulfosarcinales* showed that the ANME-2 cells were found in the proximity to *Desulfosarcinales* cells (Figure 5.4D).

We were unable to test ANME-1 by CARD-FISH from the BTF enriched biomass due to problem with probes. However, the cocci shaped and rod shaped ANME cells (Cy3-labelled ANME-1 350, ANME-2 538, ANME-3 1249) in the enriched biomass were observed by FISH. Each rod shaped cells were around 1.5 to 2 μm in diameter (Figure 5.5C). We conclude that these cells are likely ANME-1 since no other ANME types were detected by phylogenetic analysis and only ANME-1 is rod shaped. Other cocci cells could be ANME-2 as the almost 10% of the 16s rRNA gene sequences were affiliated to ANME-2 clade, while ANME-3 was not retrieved from sequence analysis.

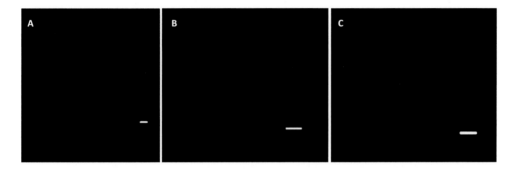

Figure 5.5 FISH images of microbial cells obtained at the end of BTF enrichment of Ginsburg Mud Volcano sediment: (A) archaeal cells hybridized by the Arc-FAM probe are in cyan blue (B and C) ANME-1(rod shaped) and ANME-2 (cocci shaped) cells hybridized with ANME-mix probe are in red. Scale bar refers to 2 μm.

5.4   Discussion

5.4.1 Anaerobic oxidation of methane in the BTF

This study shows that AOM-SR biomass was enriched in the BTF inoculated with Ginsburg mud volcano sediment as depicted by the activity of sulfate reduction with simultaneous total dissolved sulfide production after about 150 days of BTF operation (Figure 5.2B). The methane consumption gradually increased only after 120 days, and after 130 days the total dissolved sulfide production and the sulfate reduction also increased exponentially. This start-up phase was required to recover and activate the biomass for AOM-SR activity from the Ginsburg mud volcano sediment, due to prolonged (four years) storage period of the sediment at 4°C. The amount of methane consumed was almost similar to the amount of total sulfide produced/sulfate reduced as described by the stoichiometry (Eq. 5.1), i.e., a 1:1 ratio of AOM (Holler et al., 2009; Meulepas et al., 2009a; Nauhaus et al., 2002). However, the carbon dioxide production in the BTF was almost 2-3 times lower than the stoichiometric amount of methane oxidized according to Eq. 5.1. The majority of the carbon dioxide remained as bicarbonate around pH 8. In the natural environment as well, carbonate precipitation and formation of carbonate chimneys are observed very often in ANME habitats (Bayon et al., 2013; Michaelis et al., 2002; Treude et al., 2005). Another possible reason for the comparatively low carbon dioxide production could be the utilization of carbon dioxide by ANME and/or other microbes that were concomitantly enriched in the BTF. The utilization of carbon dioxide by ANME populations, especially ANME-1, has been often discussed for several marine sediments, such as the consumption of carbon dioxide by ANME-1 from Black Sea microbial mats (Treude et al., 2007) and the autotrophic mode of AOM by ANME-1 from the hydrothermal vent in Guaymas basin (Holler et al., 2011; Kellermann et al., 2012).

BTF are widely used for the treatment of industrial waste gases containing volatile organic and inorganic pollutants owing to their good mass transfer characteristics and their ability to demonstrate high removal efficiencies of pollutants at high gas flow rates (Guerrero & Bevilaqua, 2015; Kennes & Veiga, 2013; Niu et al., 2014). However, this is the first study that describes the application of a BTF for the enrichment of ANME and AOM-SR. The reactor start-up period was 120 days, which is comparatively longer than other BTF reactors used for waste gas treatment, i.e., ~ 1-2 weeks for BTF used to treat methane emissions under aerobic conditions (Avalos Ramirez et al., 2012; Estrada et al., 2014a; Estrada et al., 2014b). Nevertheless, the retention of an active AOM-SR after 120 days of BTF operation is still ~ 3 to 6 times higher compared to the AOM enrichment in a membrane bioreactor (Meulepas et al., 2009a) and hanging sponge reactor (Aoki et al., 2014) operated under ambient pressure conditions.

It is noteworthy to mention that, so far, BTF have never been tested for AOM and ANME enrichment. Aoki et al. (2014) used hanging polyurethane foam to retain the biomass within the bioreactor. This system required a start-up period of ~ 1 year. In the work of Aoki et al. (2014), the reactor volume was ~ 10 times larger than the BTF used in this study and the authors reported frequent episodes of oxygen intrusion in the reactor, which might have elongated the start-up period of that hanging sponge

164

reactor with polyurethane foam. Similar to this BTF, the hanging sponge bioreactor in Aoki et al. (2014) was incubated at ambient conditions. However, the operation was slightly different as the liquid phase was recirculated in the BTF while liquid was continuously extracted from the hanging sponge bioreactor in Aoki et al. (2014). An efficient mass transfer and biomass retention was also observed in the BTF tested for the treatment of volatile organic compounds (Pérez et al., 2016; Santos et al., 2015). Thus, a BTF with biomass retention and closed recirculation system is an appropriate system for the growth of slow growing microorganisms like ANME.

The AOM-SR rate in the BTF was in the range of 0.4 to 1.3 mM d^{-1} when methane was supplied, which is close to the value obtained by Eckernförde bay enrichment after 600 days of reactor operation (Meulepas et al., 2009b). Interestingly, sulfate reduction was observed even with intermittent supply of nitrogen instead of methane. This might be due to methane production by methanogens present in the BTF biomass, such as *Methanococcoides*, *Methanosaeta* and *Methanosarcinales* were retrieved during archaeal 16S rRNA gene analysis and the simultaneous utilization by ANME. Therefore, a small amount of methane was always detected in the BTF outlet even when methane supply was interrupted. Besides, methane was accumulated in the used polyurethane foam as packing material was inert and highly porous and presumably ANME were utilizing the residual methane stored in the polyurethane foam matrix during the nitrogen supply phases. Furthermore, the methane residence time in BTF assessed in empty bed conditions (200 mins) could be an underestimate as the actual residence time increases with the formation of biofilm in the sponges (Barcón et al., 2015; Chen et al., 2016; Mendoza et al., 2004).

Previous studies have indicated that ANME-1 biomass could be possibly derived from carbon dioxide metabolism (Kellermann et al., 2012; Treude et al., 2007; Wegener et al., 2016a; Wegener et al., 2016b). Thus, ANME-1 are regarded as methane-oxidizing chemoorganoautotrophs (Kellermann et al., 2012). Furthermore, genetic studies have revealed that ANME-1 contain genes encoding the carbon dioxide fixation pathway characteristic for methanogens (Meyerdierks et al., 2010). ANME-1 was recently proposed to be a methane producer and methane consumer. For instance, ANME-1 from the Black Sea and from the Gulf of Mexico methane seeps can produce methane from carbon dioxide (Orcutt et al., 2005; Treude et al., 2007) or methanol (Bertram et al., 2013). Thus, the AOM-SR during nitrogen supply could be due to the utilization of bicarbonate by the enriched ANME. Further research on the utilization of different carbon compounds by enriched ANME should be performed using ^{13}C labeled carbon and ANME cells and their activities can be tracked by different methods such as NanoSIMS (Musat et al., 2012) or MAR-FISH (Wagner et al., 2006).

5.4.2 ANME enrichment in BTF

This study clearly showed that the AOM-SR in the BTF was mainly performed by ANME-1 and ANME-2, with a majority of ANME-1b. These results are in contrast to the results reported from other seep sediments (Hinrichs et al., 2000; Orphan et al., 2002; Vigneron et al., 2013b) and AOM studies performed in continuous bioreactors where mainly ANME-2 was enriched (Meulepas et al., 2009a; Timmers et al., 2015; Zhang et al., 2011). Ascertaining non-detectable ANME in the initial inoculum and almost 40% of ANME-1 in the reactor after BTF operation clearly demonstrates that the designed BTF is favorable for ANME-1 growth. Although the long-term enrichment of ANME-1 has not yet been reported, the doubling time of ANME-1 has been estimated to be between 1 and 2 months (Girguis et al., 2005; Holler et al., 2011; Krüger et al., 2008). Therefore, the increasing AOM-SR activity in the BTF with a relatively shorter start-up period in comparison to other continuous bioreactors performing AOM and the microbial results evidencing the dominance of ANME-1 among the archaeal sequences show that the BTF reactor configuration is the best candidate for ANME-1 enrichment.

The highly porous polyurethane foam based packing material used in this study resembles the carbonate chimneys, which are often present in ANME habitats such as Black Sea carbonate chimneys with the predominance of ANME-1 (Blumenberg et al., 2004; Michaelis et al., 2002) and carbonate nodules (Marlow et al., 2014). The carbonate rich minerals, onto which ANME cells attach, form microbial reefs that are very porous and can thus harbour aggregates of AOM performing consortia (Marlow et al., 2014). Similar to the natural system of carbonate chimneys, the porous polyurethane foam matrix in the BTF was contributing to the proliferation of ANME-1 and high rate of AOM induced sulfate reduction.

Unlike other AOM studies, a low abundance of sulfate reducing bacteria associated with *Deltaproteobacteria* was evidenced by both sequence analysis (Figure 5.3) and CARD-FISH (Figure 5.4). This suggests a rather independent AOM in the BTF biomass. Most of the ANME-2 cells visualized were found in the vicinity of *Desulfosarcinales* cells (Figure 5.4), perhaps these sulfate reducers have a symbiotic relation with ANME-2 rather than with ANME-1. The archaeal cells were, nevertheless, not very abundantly identified by the method used in this study and we were unable to perform CARD-FISH for ANME-1 which was the major archaeal community in BTF. Therefore, it is difficult to conclude the sulfate reducing partner for the ANME enriched in this study. Typically, ANME-1 and ANME-2 associate with *Desulfosarcinales* type of sulfate reducing bacteria, and they are often found together in natural habitats and AOM enrichments (Holler et al., 2011; Knittel & Boetius, 2009; Ruff et al., 2015; Trembath-Reichert et al., 2016). However, only single shell ANME types have been observed in some habitats and microbial mats, such as in the pink microbial mat from the Black Sea containing 90% of ANME-1 (Michaelis et al., 2002) and ANME-1 in metalliferous hydrothermal vents (Wankel et al., 2012). Mostly, ANME-1 *Archaea* exist as single cells or as monospecific chains without any attached partner (Maignien et al., 2013; Orphan et al., 2002). Likewise to this study, AOM with sulfate reduction occurring in high temperature sediments is also dominated by ANME-1, suggesting the decoupling of the AOM from the sulfate reducing bacteria (Wankel et al., 2012).

5.5 Conclusions

In conclusion, by considering the key strategy of biomass retention and continuous supply of methane, the BTF was successfully applied to study AOM and enrich ANME. This study has achieved a quick start-up period of 120 days for AOM-SR in comparison to around 400 days starting period for previous bioreactors operated at ambient pressure. High rate of AOM induced sulfate reduction (with methane consumption rates of around 1 mM d^{-1}) was obtained during the active phase of BTF operation, which was due to the proliferation of ANME in the polyurethane foam packing material. The ANME-1 clade was enriched in the BTF with the majority of ANME-1b, which might be favored by the configuration and methane loading of the BTF.

5.6 References

Aoki, M., Ehara, M., Saito, Y., Yoshioka, H., Miyazaki, M., Saito, Y., Miyashita, A., Kawakami, S., Yamaguchi, T., Ohashi, A., Nunoura, T., Takai, K., Imachi, H. 2014. A long-term cultivation of an anaerobic methane-oxidizing microbial community from deep-sea methane-seep sediment using a continuous-flow bioreactor. *PLoS ONE*, **9**(8), e105356.

Avalos Ramirez, A., Jones, J.P., Heitz, M. 2012. Methane treatment in biotrickling filters packed with inert materials in presence of a non-ionic surfactant. *J. Chem. Technol. Biotechnol.*, **87**(6), 848-853.

Barcón, T., Hernández, J., Gómez-Cuervo, S., Garrido, J.M., Omil, F. 2015. Characterization and biological abatement of diffuse methane emissions and odour in an innovative wastewater treatment plant. *Environ. Technol.*, **36**(16), 2105-2114.

Bayon, G., Dupre, S., Ponzevera, E., Etoubleau, J., Cheron, S., Pierre, C., Mascle, J., Boetius, A., de Lange, G.J. 2013. Formation of carbonate chimneys in the Mediterranean Sea linked to deep-water oxygen depletion. *Nat. Geosci.*, **6**(9), 755-760.

Bertram, S., Blumenberg, M., Michaelis, W., Siegert, M., Krüger, M., Seifert, R. 2013. Methanogenic capabilities of ANME-archaea deduced from ^{13}C-labelling approaches. *Environ. Microbiol.*, **15**(8), 2384-2393.

Bhattarai, S., Cassarini, C., Gonzalez-Gil, G., Egger, M., Slomp, C.P., Zhang, Y., Esposito, G., Lens, P.N.L. 2017. Anaerobic methane-oxidizing microbial community in a coastal marine sediment: anaerobic methanotrophy dominated by ANME-3. *Microb. Ecol.*, **74**(3), 608-622.

Blumenberg, M., Seifert, R., Reitner, J., Pape, T., Michaelis, W. 2004. Membrane lipid patterns typify distinct anaerobic methanotrophic consortia. *Proc. Natl. Acad. Sci. USA*, **101**(30), 11111-11116.

Boetius, A., Ravenschlag, K., Schubert, C.J., Rickert, D., Widdel, F., Gieseke, A., Amann, R., Jorgensen, B.B., Witte, U., Pfannkuche, O. 2000. A marine microbial consortium apparently mediating anaerobic oxidation of methane. *Nature*, **407**(6804), 623-626.

Caporaso, J.G., Lauber, C.L., Walters, W.A., Berg-Lyons, D., Huntley, J., Fierer, N., Owens, S.M., Betley, J., Fraser, L., Bauer, M. 2012. Ultra-high-throughput microbial community analysis on the Illumina HiSeq and MiSeq platforms. *ISME J.*, **6**(8), 1621-1624.

Chen, Y., Wang, X., He, S., Zhu, S., Shen, S. 2016. The performance of a two-layer biotrickling filter filled with new mixed packing materials for the removal of H_2S from air. *J. Environ. Manage.*, **165**, 11-16.

Daims, H., Brühl, A., Amann, R., Schleifer, K.-H., and Wagner, M. 1999. The domain-specific probe EUB338 is insufficient for the detection of all *Bacteria*: development and evaluation of a more comprehensive probe set. *Syst. Appl. Microbiol.*, **22**(3), 434-444.

DeLong, E.F. 1992. Archaea in coastal marine environments. *Proc. Natl. Acad. Sci. USA*, **89**(12), 5685-5689.

Deusner, C., Meyer, V., Ferdelman, T. 2009. High-pressure systems for gas-phase free continuous incubation of enriched marine microbial communities performing anaerobic oxidation of methane. *Biotechnol. Bioeng.*, **105**(3), 524-533.

Estrada, J.M., Dudek, A., Muñoz, R., Quijano, G. 2014a. Fundamental study on gas-liquid mass transfer in a biotrickling filter packed with polyurethane foam. *J. Chem. Technol. Biotechnol.*, **89**(9), 1419-1424.

Estrada, J.M., Lebrero, R., Quijano, G., Pérez, R., Figueroa-González, I., García-Encina, P.A., Muñoz, R. 2014b. Methane abatement in a gas-recycling biotrickling filter: evaluating innovative operational strategies to overcome mass transfer limitations. *Chem. Eng. J.*, **253**, 385-393.

Girguis, P.R., Cozen, A.E., DeLong, E.F. 2005. Growth and population dynamics of anaerobic methane-oxidizing archaea and sulfate-reducing bacteria in a continuous-flow bioreactor. *Appl. Environ. Microbiol.*, **71**(7), 3725-3733.

Gray, N.D., Miskin, I.P., Kornilova, O., Curtis, T.P., Head, I.M. 2002. Occurrence and activity of *Archaea* in aerated activated sludge wastewater treatment plants. *Environ. Microbiol.*, **4**(3), 158-168.

Guerrero, R.B., Bevilaqua, D. 2015. Biotrickling filtration of biogas produced from the wastewater treatment plant of a brewery. *J. Environ. Eng.*, **141**(8), 04015010.

He, Y., Li, M., Perumal, V., Feng, X., Fang, J., Xie, J., Sievert, S., Wang, F. 2016. Genomic and enzymatic evidence for acetogenesis among multiple lineages of the archaeal phylum Bathyarchaeota widespread in marine sediments. *Nat. Microbiol.*, **1**, 16035.

Hinrichs, K.-U., Summons, R.E., Orphan, V.J., Sylva, S.P., Hayes, J.M. 2000. Molecular and isotopic analysis of anaerobic methane-oxidizing communities in marine sediments. *Org. Geochem.*, **31**(12), 1685-1701.

Hinrichs, K.-U., Hayes, J.M., Sylva, S.P., Brewer, P.G., DeLong, E.F. 1999. Methane-consuming archaebacteria in marine sediments. *Nature*, **398**(6730), 802-805.

Holler, T., Wegener, G., Knittel, K., Boetius, A., Brunner, B., Kuypers, M., M. M., Widdel, F. 2009. Substantial $^{13}C/^{12}C$ and D/H fractionation during anaerobic oxidation of methane by marine consortia enriched *in vitro*. *Environ. Microbiol. Rep.*, **1**(5), 370-376.

Holler, T., Widdel, F., Knittel, K., Amann, R., Kellermann, M.Y., Hinrichs, K.-U., Teske, A., Boetius, A., Wegener, G. 2011. Thermophilic anaerobic oxidation of methane by marine microbial consortia. *ISME J.*, **5**(12), 1946-1956.

Jagersma, G.C., Meulepas, R.J.W., Heikamp-de Jong, I., Gieteling, J., Klimiuk, A., Schouten, S., Sinninghe Damsté, J.S., Lens, P.N.L., Stams, A.J. 2009. Microbial diversity and community structure of a highly active anaerobic methane-oxidizing sulfate-reducing enrichment. *Environ. Microbiol.*, **11**(12), 3223-3232.

Kellermann, M.Y., Wegener, G., Elvert, M., Yoshinaga, M.Y., Lin, Y.-S., Holler, T., Mollar, X.P., Knittel, K., Hinrichs, K.-U. 2012. Autotrophy as a predominant mode of carbon fixation in anaerobic methane-oxidizing microbial communities. *Proc. Nat. Acad. Sci. U. S. A.,* **109**(47), 19321-19326.

Kennes, C., Veiga, M.C. 2013. Bioreactors for waste gas treatment. in: Kennes, C., Veiga, M.C. (Eds.), (Vol 4), Springer Netherlands, Dodrecht, the Netherlands.

Knittel, K., Boetius, A. 2009. Anaerobic oxidation of methane: progress with an unknown process. *Annu. Rev. Microbiol.*, **63**(1), 311-334.

Knittel, K., Lösekann, T., Boetius, A., Kort, R., Amann, R. 2005. Diversity and distribution of methanotrophic archaea at cold seeps. *Appl. Environ. Microbiol.*, **71**(1), 467-479.

Krüger, M., Blumenberg, M., Kasten, S., Wieland, A., Känel, L., Klock, J.-H., Michaelis, W., Seifert, R. 2008. A novel, multi-layered methanotrophic microbial mat system growing on the sediment of the Black Sea. *Environ. Microbiol.*, **10**(8), 1934-1947.

Losekann, T., Knittel, K., Nadalig, T., Fuchs, B., Niemann, H., Boetius, A., Amann, R. 2007. Diversity and abundance of aerobic and anaerobic methane oxidizers at the Haakon Mosby mud volcano, Barents Sea. *Appl. Environ. Microbiol.*, **73**(10), 3348-3362.

Maignien, L., Parkes, R.J., Cragg, B., Niemann, H., Knittel, K., Coulon, S., Akhmetzhanov, A., Boon, N. 2013. Anaerobic oxidation of methane in hypersaline cold seep sediments. *FEMS Microbiol. Ecol.*, **83**(1), 214-231.

Marlow, J.J., Steele, J.A., Ziebis, W., Thurber, A.R., Levin, L.A., Orphan, V.J. 2014. Carbonate-hosted methanotrophy represents an unrecognized methane sink in the deep sea. *Nat. Commun.*, **5**(5094), 1-12.

Mendoza, J., Prado, O., Veiga, M.C., Kennes, C. 2004. Hydrodynamic behaviour and comparison of technologies for the removal of excess biomass in gas-phase biofilters. *Water Res.*, **38**(2), 404-413.

Meng, J., Xu, J., Qin, D., He, Y., Xiao, X., Wang, F. 2014. Genetic and functional properties of uncultivated MCG archaea assessed by metagenome and gene expression analyses. *ISME J.*, **8**(3), 650-659.

Meulepas, R.J.W., Jagersma, C.G., Gieteling, J., Buisman, C.J.N., Stams, A.J.M., Lens, P.N.L. 2009a. Enrichment of anaerobic methanotrophs in sulfate-reducing membrane bioreactors. *Biotechnol. Bioeng.*, **104**(3), 458-470.

Meulepas, R.J.W., Jagersma, C.G., Khadem, A.F., Buisman, C.J.N., Stams, A.J.M., Lens, P.N.L. 2009b. Effect of environmental conditions on sulfate reduction with methane as electron donor by an Eckernförde Bay enrichment. *Environ. Sci. Technol.*, **43**(17), 6553-6559.

Meyerdierks, A., Kube, M., Kostadinov, I., Teeling, H., Glöckner, F.O., Reinhardt, R., Amann, R. 2010. Metagenome and mRNA expression analyses of anaerobic methanotrophic archaea of the ANME-1 group. *Environ. Microbiol.*, **12**(2), 422-439.

Michaelis, W., Seifert, R., Nauhaus, K., Treude, T., Thiel, V., Blumenberg, M., Knittel, K., Gieseke, A., Peterknecht, K., Pape, T., Boetius, A., Amann, R., Jørgensen, B.B., Widdel, F., Peckmann, J., Pimenov, N.V., Gulin, M.B. 2002. Microbial reefs in the Black Sea fueled by anaerobic oxidation of methane. *Science*, **297**(5583), 1013-1015.

Musat, N., Foster, R., Vagner, T., Adam, B., Kuypers, M.M. 2012. Detecting metabolic activities in single cells, with emphasis on nanoSIMS. *FEMS Microbiol. Rev.*, **36**(2), 486-511.

Nauhaus, K., Albrecht, M., Elvert, M., Boetius, A., Widdel, F. 2007. *In vitro* cell growth of marine archaeal-bacterial consortia during anaerobic oxidation of methane with sulfate. *Environ. Microbiol.*, **9**(1), 187-196.

Nauhaus, K., Boetius, A., Kruger, M., Widdel, F. 2002. *In vitro* demonstration of anaerobic oxidation of methane coupled to sulphate reduction in sediment from a marine gas hydrate area. *Environ. Microbiol.*, **4**(5), 296-305.

Niemann, H., Duarte, J., Hensen, C., Omoregie, E., Magalhaes, V.H., Elvert, M., Pinheiro, L.M., Kopf, A., Boetius, A. 2006a. Microbial methane turnover at mud volcanoes of the Gulf of Cadiz. *Geochim. Cosmochim. Ac.*, **70**(21), 5336-5355.

Niemann, H., Losekann, T., de Beer, D., Elvert, M., Nadalig, T., Knittel, K., Amann, R., Sauter, E.J., Schluter, M., Klages, M., Foucher, J.P., Boetius, A. 2006b. Novel microbial communities of the Haakon Mosby mud volcano and their role as a methane sink. *Nature*, **443**(7113), 854-858.

Niu, H., Leung, D.Y., Wong, C., Zhang, T., Chan, M., Leung, F.C. 2014. Nitric oxide removal by wastewater bacteria in a biotrickling filter. *J. Environ. Sci.*, **26**, 555-565.

Oni, O.E., Miyatake, T., Kasten, S., Richter-Heitmann, T., Fischer, D., Wagenknecht, L., Ksenofontov, V., Kulkarni, A., Blumers, M., Shylin, S. 2015. Distinct microbial populations are tightly linked to the profile of dissolved iron in the methanic sediments of the Helgoland mud area, North Sea. *Front. Microbiol.*, **6**(365), 1-15.

Orcutt, B., Boetius, A., Elvert, M., Samarkin, V., Joye, S.B. 2005. Molecular biogeochemistry of sulfate reduction, methanogenesis and the anaerobic oxidation of methane at Gulf of Mexico cold seeps. *Geochim. Cosmochim. Ac.*, **69**(17), 4267-4281.

Orphan, V.J., Hinrichs, K.-U., Ussler, W., Paull, C.K., Taylor, L.T., Sylva, S.P., Hayes, J.M., Delong, E.F. 2001. Comparative analysis of methane-oxidizing archaea and sulfate-reducing bacteria in anoxic marine sediments. *Appl. Environ. Microbiol.*, **67**(4), 1922-1934.

Orphan, V.J., House, C.H., Hinrichs, K.-U., McKeegan, K.D., DeLong, E.F. 2002. Multiple archaeal groups mediate methane oxidation in anoxic cold seep sediments. *Proc. Nat. Acad. Sci. USA*, **99**(11), 7663-7668.

Pérez, M., Álvarez-Hornos, F., Engesser, K., Dobslaw, D., Gabaldón, C. 2016. Removal of 2-butoxyethanol gaseous emissions by biotrickling filtration packed with polyurethane foam. *New Biotechnol.*, **33**(2), 263-272.

Pernthaler, A., Pernthaler, J., and Amann, R. 2004. Sensitive multi-color fluorescence *in situ* hybridization for the identification of environmental microorganisms. in: Kowalchuk, G., de Bruijn, F., Head, I.M., Akkermans, A.D., van Elsas, J.D. (Eds.), *Molecular Microbial Ecology Manual.* Vol. 1 and 2 (2nd Edition), Springer Netherlands, Dordrecht, the Netherlands, pp. 711-725.

Pernthaler, A., Pernthaler, J. and Amann, R., 2002. Fluorescence *in situ* hybridization and catalyzed reporter deposition for the identification of marine bacteria. *Appl. Environ. Microbiol.*, **68**(6), 3094-3101.

Pruesse, E., Quast, C., Knittel, K., Fuchs, B.M., Ludwig, W., Peplies, J., Glöckner, F.O. 2007. SILVA: a comprehensive online resource for quality checked and aligned ribosomal RNA sequence data compatible with ARB. *Nucleic Acids Res.*, **35**(21), 7188-7196.

Reeburgh, W.S. 2007. Oceanic methane biogeochemistry. *Chem. Rev.*, **107**(2), 486-513.

Ruff, S.E., Biddle, J.F., Teske, A.P., Knittel, K., Boetius, A., Ramette, A. 2015. Global dispersion and local diversification of the methane seep microbiome. *Proc. Nat. Acad. Sci. USA*, **112**(13), 4015-4020.

Ruff, S.E., Kuhfuss, H., Wegener, G., Lott, C., Ramette, A., Wiedling, J., Knittel, K., Weber, M. 2016. Methane seep in shallow-water permeable sediment harbors high diversity of anaerobic methanotrophic communities, Elba, Italy. *Front. Microbiol.*, **7**(374), 1-20.

Santos, A., Guimerà, X., Dorado, A.D., Gamisans, X., Gabriel, D. 2015. Conversion of chemical scrubbers to biotrickling filters for VOCs and H_2S treatment at low contact times. *Appl. Microbiol. Biotechnol.*, **99**(1), 67-76.

Schloss, P.D., Westcott, S.L. 2011. Assessing and improving methods used in operational taxonomic unit-based approaches for 16S rRNA gene sequence analysis. *Appl. Environ. Microbiol.*, **77**(10), 3219-3226.

Schreiber, L., Holler, T., Knittel, K., Meyerdierks, A., Amann, R. 2010. Identification of the dominant sulfate-reducing bacterial partner of anaerobic methanotrophs of the ANME-2 clade. *Environ. Microbiol.*, **12**(8), 2327-2340.

Siegel, L.M. 1965. A direct microdetermination for sulfide. *Anal. Biochem.*, **11**(1), 126-132.

Snaidr, J., Amann, R., Huber, I., Ludwig, W., and Schleifer, K.H. 1997. Phylogenetic analysis and in situ identification of bacteria in activated sludge. *Appl. Environ. Microbiol.* **63**, 2884-2896.

Song, Z.Q., Wang, F.P., Zhi, X.Y., Chen, J.Q., Zhou, E.M., Liang, F., Xiao, X., Tang, S.K., Jiang, H.C., Zhang, C.L. 2013. Bacterial and archaeal diversities in Yunnan and Tibetan hot springs, China. *Environ. Microbiol.*, **15**(4), 1160-1175.

Stahl DA, Amann RI. 1991. Development and application of nucleic acid probes. in: Stackebrandt, E., Goodfellow, M. (Eds.), *Nucleic acid techniques in bacterial systematics*. John Wiley & Sons Ltd, Chichester, UK, pp 205-248.

Takai, K., Horikoshi, K. 2000. Rapid detection and quantification of members of the archaeal community by quantitative PCR using fluorogenic probes. *Appl. Environ. Microbiol.*, **66**(11), 5066-5072.

Timmers, P.H., Gieteling, J., Widjaja-Greefkes, H.A., Plugge, C.M., Stams, A.J., Lens, P.N.L., Meulepas, R.J. 2015. Growth of anaerobic methane-oxidizing archaea and sulfate-reducing bacteria in a high-pressure membrane capsule bioreactor. *Appl. Environ. Microbiol.*, **81**(4), 1286-1296.

Trembath-Reichert, E., Case, D.H., Orphan, V.J. 2016. Characterization of microbial associations with methanotrophic archaea and sulfate-reducing bacteria through statistical comparison of nested Magneto-FISH enrichments. *PeerJ*, **4**, e1913.

Treude, T., Knittel, K., Blumenberg, M., Seifert, R., Boetius, A. 2005. Subsurface microbial methanotrophic mats in the Black Sea. *Appl. Environ. Microbiol.*, **71**(10), 6375-6378.

Treude, T., Orphan, V.J., Knittel, K., Gieseke, A., House, C.H., Boetius, A. 2007. Consumption of methane and CO_2 by methanotrophic microbial mats from gas seeps of the anoxic Black Sea. *Appl. Environ. Microbiol.*, **73**(7), 2271-2283.

Vigneron, A., Cruaud, P., Pignet, P., Caprais, J.-C., Cambon-Bonavita, M.-A., Godfroy, A., Toffin, L. 2013. Archaeal and anaerobic methane oxidizer communities in the Sonora Margin cold seeps, Guaymas Basin (Gulf of California). *ISME J.*, **7**(8), 1595-1608.

Vigneron, A., L'Haridon, S., Godfroy, A., Roussel, E.G., Cragg, B.A., Parkes, R.J., Toffin, L. 2015. Evidence of active methanogen communities in shallow sediments of the Sonora Margin cold seeps. *Appl. Environ. Microbiol.*, **81**(10), 3451-3459.

Villa-Gomez D, Ababneh H, Papirio S, Rousseau D.P.L., Lens P.N.L. 2011. Effect of sulfide concentration on the location of the metal precipitates in inversed fluidized bed reactors. *J. Hazard. Mater.,* **192**(1), 200-207.

Wagner, M., Amann, R., Lemmer, H., and Schleifer, K.-H. 1993. Probing activated sludge with oligonucleotides specific for proteobacteria: inadequacy of culture-dependent methods for describing microbial community structure. *Appl. Environ. Microbiol.* **59,** 1520-1525.

Wagner, M., Nielsen, P.H., Loy, A., Nielsen, J.L., Daims, H. 2006. Linking microbial community structure with function: fluorescence in situ hybridization-microautoradiography and isotope arrays. *Curr. Opin. Biotechnol.*, **17**(1), 83-91.

Wankel, S.D., Adams, M.M., Johnston, D.T., Hansel, C.M., Joye, S.B., Girguis, P.R. 2012. Anaerobic methane oxidation in metalliferous hydrothermal sediments: influence on carbon flux and decoupling from sulfate reduction. *Environ. Microbiol.*, **14**(10), 2726-2740.

Wegener, G., Boetius, A. 2008. Short-term changes in anaerobic oxidation of methane in response to varying methane and sulfate fluxes. *Biogeosci. Discuss.*, **5**(4), 3069-3090.

Wegener, G., Kellermann, M.Y., Elvert, M. 2016a. Tracking activity and function of microorganisms by stable isotope probing of membrane lipids. *Curr. Opin. Biotechnol.*, **41**, 43-52.

Wegener, G., Krukenberg, V., Ruff, S.E., Kellermann, M.Y., Knittel, K. 2016b. Metabolic capabilities of microorganisms involved in and associated with the anaerobic oxidation of methane. *Front. Microbiol.*, **7**(46), 1-16.

Zhang, Y., Henriet, J.-P., Bursens, J., Boon, N. 2010. Stimulation of in vitro anaerobic oxidation of methane rate in a continuous high-pressure bioreactor. *Bioresour. Technol.*, **101**(9), 3132-3138.

Zhang, Y., Maignien, L., Zhao, X., Wang, F., Boon, N. 2011. Enrichment of a microbial community performing anaerobic oxidation of methane in a continuous high-pressure bioreactor. *BMC Microbiol.*, **11**(137), 1-8.

CHAPTER 6

Enrichment of ANME-2 Dominated Anaerobic Oxidation of Methane Coupled to Sulfate Reduction Consortia from Cold Seep Sediment (Ginsburg Mud Volcano, Gulf of Cadiz) in a Membrane Bioreactor

The modified version of this chapter is under revision as:

Bhattarai S., Cassarini C., Rene E.R., Kümmel S., Esposito G. and Lens P.N.L. (2018) Enrichment of ANME-2 dominated anaerobic methanotrophy from cold seep sediment in an external ultrafiltration membrane bioreactor. *Life Science Engineering,0,* 1-11.

Abstract

In this study, anaerobic methanotrophs (ANME) *Archaea* present in sediment from the Ginsburg mud volcano (Gulf of Cadiz), a known anaerobic oxidation of methane (AOM) site, was enriched in a membrane bioreactor (MBR). AOM activity coupled to sulfate reduction was monitored in the membrane bioreactor (MBR) for 726 days at 22 (± 3)°C. The MBR was equipped with a cylindrical ultrafiltration membrane, fed a defined medium containing artificial seawater and operated at a cross flow velocity of 0.02 m min^{-1}. Methane consumption was observed with simultaneous sulfate reduction at a volumetric rate of 0.7 mM of methane consumption d^{-1} and 0.5 mM of sulfate reduction d^{-1}, respectively. The enriched biomass was tested for AOM activity during the end of the MBR operation. In batch incubations, when this biomass was incubated anaerobically with ^{13}C labeled methane, ^{13}C labeled inorganic carbon was produced and the AOM rate based on ^{13}C-inorganic carbon production rate amounted to 1.2 µmol g_{dw}^{-1} d^{-1}. The incubation with ^{13}C labeled carbonate showed an even higher rate of ^{13}C-inorganic carbon production of 34.5 µmol g_{dw}^{-1} d^{-1}. Catalyzed reporter deposition - fluorescence *in situ* hybridization (CARD-FISH) analysis of the enriched biomass after ca. 400 days and 726 days, respectively, showed ANME-2 and *Desulfosarcina* type sulfate reducing bacteria were enriched in the MBR, which formed closely associated aggregates. The intermittent observation of acetate production in the bioreactor suggests that acetate produced during methane consumption might be an intermediate for electron transfer between the ANME and sulfate reducers.

6.1   Introduction

The anaerobic oxidation of methane (AOM) coupled to sulfate reduction is a prominent methane consuming microbial process occurring in anoxic marine sediments (Knittel & Boetius, 2009; Reeburgh, 2007). From a thermodynamic view point, AOM to sulfate reduction (AOM-SR) yields minimal energy: only 17 kJ mol^{-1} of energy is released under standard conditions in the overall reaction (Eq. 6.1).

$$CH_4 + SO_4^{2-} \rightarrow HCO_3^- + HS^- + H_2O \quad \Delta G° = 17 \text{ kJ mol}^{-1} CH_4 \quad \text{Eq. 6.1}$$

The AOM-SR process is common in sites with methane seeping or diffusing to the sediment surface, such as methane seeps (Boetius & Wenzhofer, 2013; Ruff et al., 2013; Ruff et al., 2016), methane hydrates (Boetius et al., 2000; Knittel et al., 2005; Treude et al., 2003), hydrothermal vents (Holler et al., 2011; Vigneron et al., 2013) and methanogenic sediments (Bhattarai et al. 2017; Treude et al., 2014; Treude et al., 2005b). In those specific niches, few phylogenetically distinct microbial clades i.e., anaerobic methanotrophs (ANME) *Archaea* and their associated sulfate reducing bacteria, are able to grow via energy obtained from AOM coupled to sulfate reduction.

Three groups of ANME have been distinguished so far, of which ANME-1 and ANME-2 are the most abundant and geographically widespread groups (Hinrichs et al., 2000; Knittel & Boetius, 2009; Orphan et al., 2001), whereas ANME-3 have been retrieved from few marine mud volcanoes (Niemann et al., 2006b). *Desulfosarcina/Desulfococcus group* and *Desulfobulbus* clades of *Deltaproteobacteria* are well known sulfate reducing partners of ANME. There is still a major knowledge gap in AOM studies about the mechanism and the physiology of ANME. Some of the studies have postulated independent AOM performed solely by ANME (Milucka et al., 2012; Scheller et al., 2016), whilst consortia of ANME and sulfate reducing bacteria may act together for AOM by interspecies electron transfer (McGlynn et al., 2015).

A major hindrance for studies pertaining to the AOM mechanism is the difficulty to culture or enrich ANME, imputable to the requirements of rigorous anaerobic conditions, constant methane availability and the extreme slow growth rate of the ANME. Although designing special bioreactors for the enrichment of ANME can be a relatively complex engineering task, problems with process stability will always exist because of their slow growth and environmental sensitiveness. Nevertheless, a relatively robust bioreactor design is required for their enrichment and cultivation.

Several AOM rate determining and ANME studies have been performed in batch and continuous bioreactors using several different ANME enrichments often with conflicting or largely varying results. For instance, AOM-SR rates were observed in a wide range of 0.375-286 μmol g_{dw}^{-1} d^{-1} in continuous systems (Aoki et al., 2014; Girguis et al., 2005; Meulepas et al., 2009), 1-20 μmol g_{dw}^{-1} d^{-1} in batch systems

(Holler et al., 2011; Kruger et al., 2008; Treude et al., 2007) and up to 230 µmol g_{dw}^{-1} d^{-1} in a fed-batch system (Nauhaus et al., 2007). In a continuous bioreactor with submerged membranes, inoculated with Eckernförde Bay sediment, the maximum AOM rate was 286 µmol g_{dw}^{-1} d^{-1}, whereas the bioreactor showed similar AOM activity only after 400 days of continuous operation (Meulepas et al., 2009). Further, the AOM rate obtained from southern Hydrate Ridge, Offshore Oregon was 230 µmol g_{dw}^{-1} d^{-1} when that sediment was incubated in a fed-batch bioreactor (Nauhaus et al., 2002). In the same fed-batch incubation system, the AOM-SR rate of the sediment from Hydrate Ridge was significantly higher at elevated methane partial pressures of 14 atm (Nauhaus et al., 2007; Nauhaus et al., 2005). On the other hand, in a batch system inoculated with sediment from the Gulf of Mexico, the AOM rate was almost 10 times lower than the previously mentioned fed-batch and continuous bioreactors (Kruger et al., 2008). Besides, almost 90% of the ANME-2d enrichment was achieved with a continuous hollow-fiber membrane reactor performing methane-dependent denitrification (Shi et al., 2013).

Previous experiences with bioreactors have shown that the continuous supply of methane and biomass retention capacity of the reactor are important engineering considerations for ANME enrichment and achieving higher AOM rates. Therefore, this study focused on ensuring a continuous supply of methane to ANME by considering the biomass retention as a key element during bioreactor operation. For this purpose, a membrane bioreactor (MBR) was inoculated by Ginsburg mud volcano sediment and operated for the study of AOM and enrich the microorganisms performing AOM. To overcome the slow growth rate of ANME, an external membrane was used in the MBR, which allows complete biomass retention. Further, the AOM rate was measured by activity assays with ^{13}C labeled methane and the results were compared with other AOM-SR bioreactors. The microbial community performing AOM was tracked by using catalyzed reporter deposition-fluorescence *in situ* hybridization (CARD-FISH).

6.2 Materials and Methods

6.2.1 Origin and storage of inoculum

Sediment samples used as the MBR inoculum were obtained from the Ginsburg mud volcano, Gulf of Cadiz, from 53 cm to 153 cm below the sea floor. The sampling location was in the crater of Ginsburg mud volcano (35° 22.431'N; 07° 05.291'W) at a water depth of ca. 910 m (Zhang et al., 2010). The sediment core was immediately capped and stored under anaerobic conditions at 4°C, as described previously by Zhang et al. (2010). Ginsburg mud volcano is one of the noted sites for AOM among the mud volcano in the Gulf of Cadiz (Niemann et al., 2006a). The microbial community in the initial inoculum was recently characterized by the high throughput sequencing of 16S rRNA genes for both *Archaea* and *Bacteria* by Illumina Miseq method (see chapter 5).

6.2.2 MBR design and mode of operation

The MBR was designed and fabricated in-house and operated for more than two years i.e. 726 days to study AOM coupled to sulfate reduction and enrich the microbial community responsible for AOM (Figure 6.1). The MBR was made of a glass column (length: 65 cm, diameter: 100 mm). A hollow fiber ultrafiltration membrane from Pentair X-Flow (Enschede, the Netherlands) having a pore size < 2 μm with a cross flow velocity of 0.02 m min^{-1} was used and placed externally in the recirculation loop. This membrane allowed the permeate to pass through while retaining the biomass in the reactor. The MBR was completely sealed gas-tight to prevent any possible leakage or air intrusion during its operation. The reactor was equipped with a gas diffuser for sparging methane, a Smart Thermal Mass Flow Controller type SLA5850 (Brooks, Veenendaal, the Netherlands), one way valves and sampling ports (Figure 6.1).

The inoculum was diluted 3 times using artificial sea water and activated in two-500 ml serum bottles with 2 bar methane pressure for 30 days prior to the inoculation in the MBR. The tubings, joints, connectors and the reactor column were cleaned with ethanol and dried under nitrogen, and then with sterilized water prior to the start of the experiments in the MBR. The MBR was left under nitrogen for one week and inoculation was performed in the presence of nitrogen.

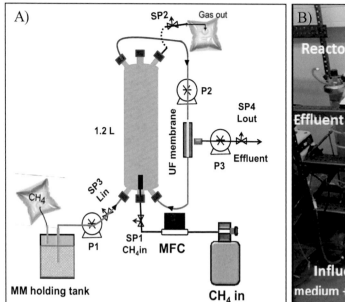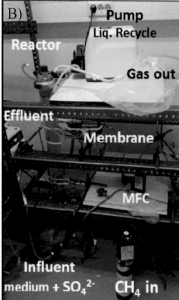

Figure 6.1 Membrane bioreactor for the anaerobic oxidation of methane and enrichment of anaerobic methanotrophs A) Schematic configuration of MBR, B) Image of MBR set-up in laboratory. Here, SP: sampling port, UF: ultrafiltration, MM: artificial seawater medium and MFC: mass flow controller. P1, P2 and P3 were the pumps used in the bioreactor. Lin and Lout represent the liquid fed (artificial sea water medium) to MBR and liquid ejected out from MBR.

After inoculation, the MBR was continuously supplied with methane and operated in sequencing batch mode for the liquid phase. The artificial seawater medium was continuously fed and recirculated in the reactor and passed from the bottom of the bioreactor to the ultrafiltration membrane and back to the reactor at a rate of 20 ml min^{-1}. The recirculation system enhances the liquid-gas contact as well as allows the biomass to grow in suspension. Under the continuous mode of liquid supply, the effluent was extracted from the ultrafiltration membrane. The turbidity of effluent tank was regularly monitored and the effluent was centrifuged at 16 g in a micro centrifuge (ThermoFisher Scientific) and also visualized under a stereo microscope (Olympus, Zoeterwoude, the Netherlands). However, no pellet was formed and cells were not observed in the effluent showing the fact that the used ultra filtration membrane effectively retained the biomass in bioreactor.

The artificial seawater medium was prepared according to the composition recommended in a previous study (Zhang et al., 2010), in per liter of demineralised water: NaCl (26 g), KCl (0.5 g) $MgCl_2.6H_2O$ (5 g), NH_4Cl (0.3 g), $CaCl_2\cdot2H_2O$ (1.4 g), Na_2SO_4 (1.43 g), KH_2PO_4 (0.1 g), trace element solution (1 ml), 1 M $NaHCO_3$ (30 ml), vitamin solution (1 ml), thiamin solution (1 ml), vitamin B_{12} solution (1 ml), 0.5 g L^{-1} resazurin solution as a redox indicator (1 ml) and 0.5 M Na_2S solution (1 ml). The pH was adjusted to 7.0 with sterile 1 M Na_2CO_3 or H_2SO_4 solution. Methane gas (99.9% CH_4, Linde gas, Schiedam, the Netherlands) stored in a gas tank was supplied to the bioreactor using a mass flow controller at a rate of 0.5 ml min^{-1}. The bioreactor was operated at atmospheric pressure and a temperature of 22 ± 3°C.

The reactor effluent was sampled twice per week for sulfate, sulfide and volatile fatty acids (VFA) analysis i.e. from the sampling port SP4. The inlet (SP1) and outlet gas (SP2) were also measured twice a week for the measurement of carbon dioxide and methane concentrations. Biomass sample for the microbial visualization was obtained on day 400 and at the end of the MBR operation. The activity assay was performed with the bioreactor biomass suspension collected during the end of MBR operation (day 726).

6.2.3 Activity assay with labeled methane

Activity tests with ^{13}C labeled methane were performed in triplicate with duplicate controls under nitrogen (without methane), whereas the activity test with ^{13}C labeled carbonate was performed in duplicate. The bottles were incubated on an orbital shaker (rotating at 100 rpm), at the operation temperature of the MBR (22 ± 3°C). After determination of the exact weight and volume of the 118 ml serum bottles, they were closed with butyl rubber stoppers and caps, and the gas phase was replaced several times with nitrogen gas and made vacuum thereafter. Subsequently, 20 ml of the sampled MBR biomass suspension, obtained at the end of the reactor operation (day 726), was transferred into the serum bottles using hypodermic needles and maintained under methane rich conditions in 1 L serum bottles. To ensure homogenous sampling, the serum bottles were gently shaken before sampling. The bioreactor suspension was diluted two times with artificial seawater. Then, the headspaces of the bottles were flushed with methane for 8 min and an estimated equivalent amount of 5% of headspace was taken out and once again filled with the same amount of ^{13}C labeled methane. For the incubation with ^{13}C labeled carbonate, almost 2% equivalent amount of carbonate was added to the batches instead of ^{13}C labeled methane.

The activity assay was performed for 50 days and sampling was done on a weekly basis. 500 µl of headspace samples was withdrawn from each batch incubation and injected in 10 ml vacuumed crimped-sealed vials. The stable carbon isotope composition of methane and carbon dioxide was determined using a Gas Chromatography - Isotope Ratio Mass Spectrometer (GC-IRMS, Agilent 7890A) fitted with a CP Pora Bond column (50 m × 0.32 mm × 0.5 µm). The oxidation oven was set to 40°C and 200-1000 µl of sample was injected into the gas chromatograph that functioned in split mode (1:10). The initial temperature of the oven was 40°C which was held constant for 120 min, and thereafter it was increased to 250°C at the rate of 20°C min^{-1} and the final temperature (250°C) was held for 10 min. Measurements were performed in triplicate and standard deviation was smaller than 0.5 δ-units. Standard gas mixture of methane and carbon dioxide were measured periodically along the entire isotope analysis. The resulting methane and carbon dioxide were analyzed for the ^{13}C isotopic enrichment (^{13}F(tn)).

In addition, almost 2 ml of liquid sample was taken out for the measurement of total dissolved sulfide, sulfate and pH. Sulfate and total dissolved sulfide were measured as mentioned in section, chemical analysis.

6.2.4 Chemical analysis

For each sample, almost 5 ml of liquid was extracted from MBR. The pH was measured with a Metrohm pH meter from Metrohm Applikon B.V. (Schiedam, the Netherlands) and a pH electrode from SenTix WTW (Amsterdam, the Netherlands). Sulfate was analyzed using an Ion Chromatograph system (Dionex-ICS-1000 with AS-DV sampler), as described previously by Villa-Gomez et al. (2011). Total dissolved sulfide was analyzed spectrophotometrically using the methylene blue method at a wavelength of 750 nm (Acree et al., 1971). One sample volume (0.5 ml) was diluted to one volume of 1 M NaOH to raise the pH in order to prevent sulfide volatilization after sampling and measured immediately. The amount of sulfide measured accounted for all cumulative dissolved sulfide species (H_2S, HS^- and S^{2-}) in the bioreactor effluent. All the measurements for the chemical analysis in the liquid form were performed in duplicates.

Methane and carbon dioxide concentrations in the gas-in and gas-out of the MBR were measured by Gas Chromatography using GC 3800 from VARIAN (Middelburg, the Netherlands). The gas chromatograph was equipped with a PORABOND Q column (25 m × 0.53 mm × 10 μm) and a thermal conductivity detector (TCD). The carrier gas was helium (15 Psi), flow rate 0.5 ml min^{-1}, the oven temperature was 25°C and the gas valve injection was 0.5 mL. Measurements were performed in duplicates and a standard deviation lower than 0.5% in chromatograph peak area was used for calculations. A standard gas mixtures of methane and carbon dioxide was measured periodically together with other samples as a method of quality assurance practice.

Acetate was measured along with other VFA in a gas chromatograph (GC 340) from VARIAN (Middelburg, the Netherlands). The gas chromatograph was equipped with a flame ionization detector (FID) and CP WAX 58 (FFAP) column having the following dimensions: 25 m × 0.32 mm × 0.2 μm. The carrier gas was helium with 25 Psi at a flow rate of 78 ml min^{-1}. The oven was set at 105°C and the FID temperature was maintained at 300°C. The injection volume was 1 μl and injected by an auto sampler which was set for duplicate measurements for each sample. Prior to the measurements, the internal standard mixture of formic/propionic acid was added. For each measurement, VFA standards with known concentrations were measured in order to ensure the quality.

6.2.5 Calculations

The volumetric sulfate consumption and the total dissolved sulfide production rates from the chemical activity during MBR operation were calculated as described in Eq. 6.2 and 6.3 (Meulepas et al. 2009):

$$\text{Volumetric sulfate reduction rate (mM d}^{-1}) = \frac{[SO_4^{2-}{}_{(t)}] - [SO_4^{2-}{}_{(t+\Delta t)}]}{\Delta t} \qquad \text{Eq. 6.2}$$

$$\text{Volumetric sulfide production rate (mM d}^{-1}) = \frac{S_{(t+\Delta t)}^{2-} - S_t^{2-}}{\Delta t} \qquad \text{Eq. 6.3}$$

where, $SO_4^{2-}{}_{(t)}$ is the concentration (mM) of sulfate (SO_4^{2-}) at time (t) in mM during the batch incubation, $SO_4^{2-}{}_{(t+\Delta t)}$ is the concentration (mM) of SO_4^{2-} time (t+Δt) $S_{(t)}^{2-}$ is total sulfide concentration at time (t) and $S^{2-}_{(t+\Delta t)}$ is the concentration (mM) of SO_4^{2-} time (t+Δt). The sulfate concentration of the maximum gradient in the slope of the activity test was considered for the maximum volumetric rate calculations.

The methane oxidation rate was estimated on the basis of total dissolved inorganic carbon (DIC) produced during the activity assay incubation, as described recently by Scheller et al. (2016). The amount of DIC formed at time *n* was indicated as Δ[DIC](tn) and calculated from the measured ^{13}F (fractional abundance of ^{13}C) by neglecting the isotope effects on AOM (Eq. 6.4).

$$\Delta[DIC](tn)\text{in mM} = [DIC](t0) \times \frac{13F(tn) - 13F(t0)}{(13F(CH_4) - 13F(tn))} \qquad \text{Eq. 6.4}$$

where, [DIC] is the sum of carbonate, bicarbonate and CO_2 in mM and [DIC](t0) is the initial concentration of the inorganic carbon in the incubations i.e. 30 mM, 13F(tn)is the ^{13}F fraction of DIC at time n of incubation, 13F(t0) is the ^{13}F fraction of DIC at time 0 of incubation and $13F(CH_4)$ is the ^{13}C fraction of methane at time 0.

For each incubation, the methane oxidation rate per volume of sediment slurry was estimated by linear regression with 95% confidence intervals and the estimated Δ[DIC](tn) was plotted as derived from the regression equation. Then total amount of DIC in µmol was estimated as;

$$\text{Amount of DIC per vial (µmol)} = \Delta[DIC](tn) \times 60 \text{ ml} \qquad \text{Eq. 6.5}$$

The AOM rate on the basis of DIC production (µmol d^{-1}) was obtained from the slope gradient of the plot of Δ[DIC](tn) versus time and then estimated on the basis of gram dry weight of biomass.

6.2.6 Microbial analysis

Microbial analysis of MBR biomass was performed mainly by CARD-FISH. Prior to microbial visualization by CARD-FISH, the biomass was screened with FISH by using three different ANME specific probes, i.e. ANME-1 350 (Boetius et al., 2000), ANME-2 538 (Treude et al., 2005a), ANME-3 1249 (Niemann et al., 2006b) and fluorescence was only obtained from the ANME-2 specific probe (data not shown). Therefore, CARD-FISH was only analyzed for the ANME-2 clade and its associate sulfate reducing partner (*Desulfosarcina*). Enriched biomass from the MBR after 400 days of operation and at the end of reactor operation (726 days) were analyzed by CARD-FISH. For the biomass obtained on day 726, the biomass for CARD-FISH was taken from batch incubations upon completion of the activity assay with ^{13}C labeled methane and carbonate.

Cells were fixed in 1% paraformaldehyde (PFA) and phosphate buffer saline (PBS) overnight at 4°C. After fixation, the cells were washed three times in PBS and stored in 50% ethanol/PBS (*v/v*). Samples were filtered through a 0.22 μm pore sized polycarbonate filters (GTTP type, Millipore), supported by cellulose acetate filters (0.45 μm pore size, HAWP type, Millipore). The polycarbonate filters were stored at - 20°C.

Further, the filters were dipped in low-gelling-point agarose (0.1% *w/v* in Milli-Q water), dried on a petri dish at 37°C and dehydrated in 96% (*v/v*) ethanol. For bacterial cell wall permeabilization, samples were treated with 10 mg ml^{-1} of lysozyme, 50 mM EDTA and 0.1 M Tris-HCl, at 37°C for 1 h, washed with MilliQ water and subsequently treated with achromopeptidase (60 U ml^{-1} in 0.01 M NaCl and 0.01 M Tris-HCl, pH 8). The polycarbonated filters were washed 3 times in MilliQ water. For archaeal cell wall permeabilization, the polycarbonated filters were incubated with sodium dodecyl sulfate (SDS) and proteinase K solution (Holler et al. 2011). For all the samples, endogenous peroxidases were inactivated by incubation in 0.1% H_2O_2 for 1-2 min at room temperature. Thereafter, the filters were washed twice in MilliQ water and dehydrated with 50, 80 and 96% (*v/v*) ethanol (1 min each).

CARD-FISH with horseradish peroxidase (HRP)-labeled oligonucleotide probes and tyramide signal amplification was done according to the protocol described in the literatures (Pernthaler et al., 2002; Pernthaler et al., 2004). The microorganisms were visualized using archaeal and bacterial HRP-labeled oligonucleotide probes ARCH915 (Stahl & Amann, 1991) and EUB338-I-III (Daims et al., 1999), respectively, at a formamide concentration of 35% (*v/v* hybridization buffer). The probes DSS658 (Boetius et al., 2000) and ANME-2 538 (Schreiber et al., 2010) were used for the detection of *Desulfosarcina-Desulfococcus sp.* and ANME-2 *Archaea*, respectively, using a formamide concentration of 50%. Oligonucleotide probes were purchased from Biomers (Ulm, Germany).

For dual-CARD-FISH, peroxidases of initial hybridizations were inactivated according to the procedure described by Holler et al. (2011). Tyramide amplification was performed using the fluorochromes Alexa Fluor 488 and Alexa Fluor 594, prepared as described by Pernthaler et al. (2004). Finally, all living cells were stained with 4', 6'-diamidino-2-phenylindole (DAPI) and analyzed using an epifluorescence microscope of Carl Zeiss (Oberkochen, Germany).

6.3 Results

6.3.1 AOM in the membrane reactor

During the operation period of MBR, anaerobic conditions were maintained the MBR as the rezasurin containing reactor mixed liqour remained colorless. The pH was maintained between 7.0 and 8.0 during the reactor operation by manual addition of sulfuric acid when the pH was above 8.0, especially after 300 days. After 600 days of MBR operation, the pH increment was rapid and reached up to 8.3 (Figure 6.2A). These pH changes in the bioreactor mixed liquor are typical for AOM activity as shown by Eq. 6.1.

The first 150 days of MBR operation were the MBR start-up period. The microbial community was in its acclimatization phase and the following observations were made: sulfide productions were below 0.5 mM, the reactor showed a non-steady trend of sulfate consumption and methane consumption was not detected (Figure 6.2). After 200 days of operation, the sulfate consumption increased in MBR. Between 350 to 400 days with the sulfate reduction was active simultaneous production of sulfide, which was at nearly an equimolar ratio for a short period (~ day 370) as defined by the AOM stoichiometry (Eq. 6.1) (Figure 6.2B).

The maximum volumetric sulfate reduction rate was observed between days 360 to 380 with a rate of almost 0.5 mM of sulfate reduction d^{-1}, whereas during the same period, the sulfide production rate was ~ 0.2 mM of sulfide production d^{-1}. Another sulfide production peak was observed between days 550 and 585, with a maximum sulfide production rate of 0.1 mM of sulfide production d^{-1}. A maximum sulfide concentration of 5.5 mM and minimum sulfate concentration of 1.5 mM in the MBR were reached at day 585, with almost 11 mM sulfate consumption in 160 days.

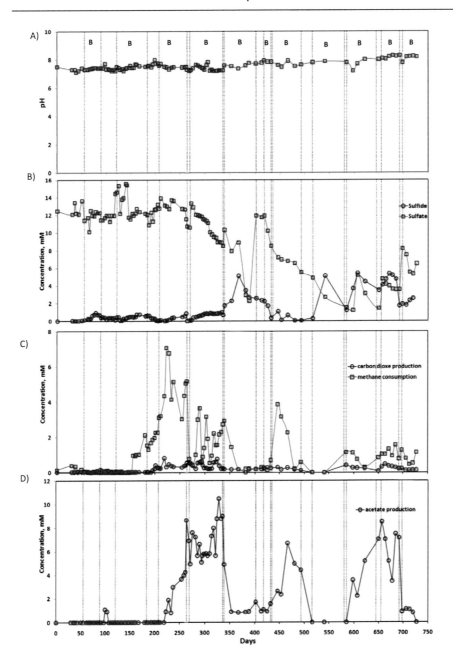

Figure 6.2 *Chemical parameters during MBR operation A) pH and effluent concentrations of B) sulfate and sulfide C) methane consumption and carbon dioxide production and D) acetate formation. Here in the figure B refers to the batch mode operation. pH, sulfate, sulfate and acetate concentration were measured from the effluent whereas, the gas data presented here is the differences between inlet gas measurement and outlet gas measurement.*

Methane consumption showed a rather fluctuating trend throughout the MBR operation, with methane consumption peaks of 2 to 7 mM during different operational periods. Methane consumption was almost undetectable during the initial 150 days of reactor operation. Thereafter, methane consumption increased up to 7 mM and a rate of 0.7 mM of methane consumption day^{-1} was observed between days 200 and 223 (Figure 6.2C). However, during that period, sulfate consumption and simultaneous sulfide production was almost 10 times lower than the methane consumption. Similarly, the carbon dioxide production was 5 to 20 times lower than the methane consumption. Furthermore, acetate production was observed intermittently during different periods of MBR operation, especially between days 200 and 520 (Figure 6.2D). The acetate production trend followed a similar trend as that of methane consumption, and the acetate production increased whenever there was an increase in the methane consumption. In some instances, acetate production was even higher than the methane consumption as it reached up to 10.5 mM on day 329, whereas the methane consumption was only 2 mM on the same day.

6.3.2 Methane oxidation and sulfate reduction rates from batch activity assays

^{13}C labeled methane was consumed by the microbial community from the MBR suspension as shown by the simultaneous ^{13}C DIC and sulfide production (Figure 6.3). However, the sulfate reduction and sulfide production were almost 20 times higher than the ^{13}C DIC production (Figures 6.3A, 6.3B and 6.3C). In control incubations with nitrogen instead of methane in the headspace, no DIC production was observed, suggesting the ^{13}C DIC production was solely by the ^{13}C methane consumption. Further, the sulfate consumption was below 50 µmol of sulfate in case of control throughout the incubation period, whereas the cumulative consumption of sulfate for the incubation with ^{13}C methane was 200 µmol of sulfate. The cumulative sulfide production was below 30 µmol throughout the control experiment, whereas in the case of incubation with ^{13}C methane, the cumulative sulfide production reached values between 4 and 5 mM. On the basis of the total DIC production, the average estimated AOM rate for the incubation with ^{13}C methane was 1.2 µmol g_{dw}^{-1} d^{-1}. Further, the sulfate reduction rate and simultaneous production rate was estimated as 29.5 µmol g_{dw}^{-1} d^{-1} and 40.5 µmol g_{dw}^{-1} d^{-1}, respectively, in the incubations with ^{13}C methane. The activity assay with ^{13}C labeled carbonate showed a higher and rapid

DIC production in comparison to the incubation with ^{13}C methane (Figure 6.3D). The estimated DIC production reached up to almost 300 µmol with a similar amount of sulfate reduction during the incubation period, whereas the DIC production of the control incubation under nitrogen was much lower and did not show an increasing trend (Figures 6.3D and 6.3E).

The total DIC production and sulfate consumption was almost in an equimolar amount; however the cumulative sulfide production was slightly higher than the total amount of sulfate consumed and reached almost 350 µmol of sulfate in case of the ^{13}C carbonate incubations (Figures 6.3C, 6.3D and 6.3E). The estimated DIC production rate in the incubation with ^{13}C bicarbonate amounted to 34.5 µmol g_{dw}^{-1} d^{-1} with a simultaneous sulfate reduction and sulfide production rate of, respectively, 26 µmol g_{dw}^{-1} d^{-1} and 48 µmol g_{dw}^{-1} d^{-1}.

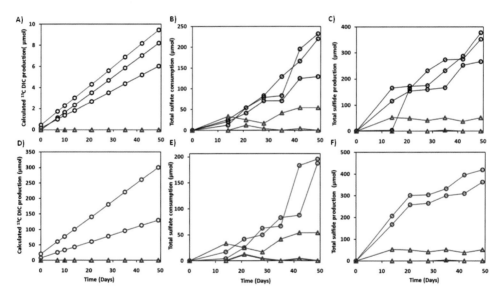

Figure 6.3 Estimated ^{13}C DIC production, sulfate consumption and sulfide production profiles during batch activity assays in 118 mL serum bottles inoculated with the MBR suspension at the end of bioreactor operation (726 days) with 5% [13]C methane (A, B and C) and 2% [13]C carbonate (D, E and F). Blue lines in the upper panel of figures (A, B and C) were the results obtained from incubations with ^{13}C methane, whereas the red lines represent the control incubations without methane or with nitrogen. Green lines in the lower panel of figures (D, E and F) were the results obtained from incubations with ^{13}C carbonate and the red lines represents the controls with nitrogen.

6.3.3 CARD-FISH analysis of the microbial community

ANME-2 cells were observed by CARD-FISH in both biomass samples, i.e. after 400 and 726 days (Figures 6.4 and 6.5). For the MBR biomass obtained on day 400, the ANME-2 cells were almost 5 to 30% of the total cells stained by DAPI (Figures 6.4A, 6.4B, 6.4C and 6.4D). In most of the microscopic observations, the total amount of ANME-2 cells was almost 30% of the total cells stained by DAPI in the biomass obtained on day 400 (Figures 6.4B and 6.4D). Nevertheless, as seen in Figure 6.4C, the ANME-2 cells were less than 5% in some of the observations after day 400 of MBR operation. The observed ANME-2 cells were cocci shaped with ~ 1 to 1.5 μm in size. Some of those ANME-2 cells appeared as circular discs with no fluorescence signal from the center of the cell (Figure 6.4B).

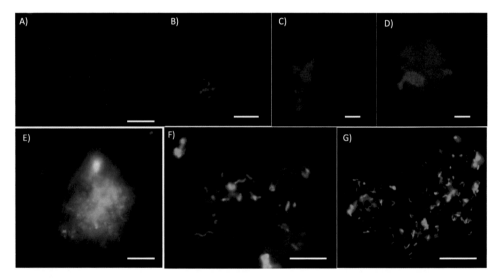

Figure 6.4 *Cell aggregates of all the microbial cells, ANME-2 and Desulfosarcinales visualised with respective fluorescent-labeled oligonucleotide probes by CARD-FISH of biomass enriched after 400 days of MBR operation. Each aggregate shown were taken from different pictures. Upper panel photomicrographs (A-D) represents large cluster of ANME-2 hybridized with ANME-2 538 probes (red) and all cells stained by DAPI (blue). Lower panel photomicrographs (E-G) represents cluster of Desulfosarcinales hybridized with DSS 658 (green) and all cells stained by DAPI (blue). The white scale bar represents 10 μm.*

Desulfosarcinales cells, a common sulfate reducing partner of ANME-2 were also visualized along with other cells that were stained by DAPI (Figures 6.4E and 6.4F). Most of these *Desulfosarcinales* were rod shaped cells with a size of ~ 1 μm. Nearly

20-30% of the total cells were estimated to be *Desulfosarcinales* from microscopic observation. It is also noteworthy mentioning that, in this study, the double hybridization for ANME-2 and *Desulfosarcinales* was not performed for the biomass obtained on day 400.

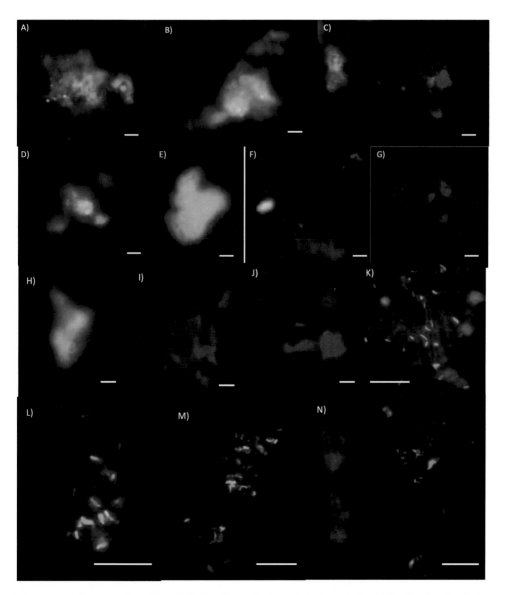

Figure 6.5 *Cell aggregates of ANME-2, Desulfosarcinales, Bacteria and microbial cells, visualized with respective fluorescent-labeled oligonucleotide probes by CARD-FISH of biomass obtained at the end of the batch activity assays (50 days), which were performed with MBR biomass on day 726. Photomicrographs (A-F and L-N) were obtained with ^{13}C methane activity assay, in which (A-F)*

represents the tight consortia of ANME-2 hybridized with ANME-2 538 (red) and Desulfosarcina hybridized with DSS 658 (green). (L-N) represents the cell clusters of Desulfosarcinales hybridized with DSS 658 (green) and other bacterial cells hybridized with EUB338-I-III (violet). (G-K) photomicrographs were obtained at the end of the incubations with ^{13}C carbonate activity assay, in which (H) represents the tight consortia of ANME-2 hybridized with ANME-2 538 probes (red) and Desulfosarcina hybridized with DSS 658 (green), (I-J) represents large cluster of ANME-2 hybridized with ANME-2 538 probes (red) and all cells stained by DAPI (blue) and (K) represents clusters of Desulfosarcinales hybridized with DSS 658 (green) and all cells stained by DAPI (blue). The white scale bar represents 10 µm.

For the biomass obtained at the end of bioreactor operation, i.e. day 726, aggregates of ANME-2 cells and Desulfosarcina were observed (Figures. 6.5A-6.5H). In most cases, ANME-2 and Desulfosarcina were clustered together with almost 1:2 ratios for ANME-2 and Desulfosarcina cells (Figures 6.5B-6.5F). However, in some observations, the amount of ANME-2 and Desulfosarcinales cells were almost similar (Figures 6.5A and 6.5H). The amount of ANME-2 cells compared to the total amounts of cells stained by DAPI was almost 20-30%. ANME-2 cells were observed for the incubations with ^{13}C methane and ^{13}C carbonate, respectively (Figures 6.5A-6.5J). However, ANME-2 cells were not clearly distinguished in the control with nitrogen atmosphere incubated for 50 days. The rod shaped Desulfosarcinales cells were detected mostly in conjunction with ANME-2 cells and also with other bacterial cells (Figures 6.5K-6.5N). In general, the amount of Desulfosarcinales cells was in a 1:1 ratio with other bacterial cells, as shown in Figures 6.5L-6.5N.

6.4　Discussion

6.4.1 AOM and sulfate reduction in the MBR

This study showed that AOM activity coupled to sulfate reduction in a MBR inoculated with the Ginsburg mud volcano sediment. The AOM and sulfate reduction rate increased after 150 days of reactor operation, showing the biomass was initially regaining its activity after a long storage period at 4°C and then the sulfate reduction increased exponentially. The use of an external ultrafiltration membrane in the MBR for the retention of slow growing microorganisms such as ANME is a promising cultivation approach for AOM studies.

The AOM activity and rate in the MBR is in the range of seep and mud volcano samples (Knittel & Boetius, 2009). However, it was lower than the so far reported rates for high pressure incubations and bioreactor systems (Deusner et al., 2009; Timmers et al., 2015; Zhang et al., 2010). As methane is poorly soluble in marine water at ambient pressure (Thauer & Shima, 2008; Yamamoto et al., 1976), the AOM reaction is highly influenced by the methane partial pressure in both natural environments and bioreactor systems. An effective and continuous methane supply to anaerobic methanotrophs is an important factor for the enrichment of ANME. Similar to a previously reported MBR, the methane was continuously bubbled in the MBR to ensure the continuous methane availability to the AOM enrichments (Meulepas et al. 2009). However the AOM rate achieved in this MBR was almost 100 times lower than the AOM rate achieved in that MBR operated by Meulepas et al. (2009). The inoculum used in the previous MBR study was from Eckernförde Bay with a water depth of 28 m (Meulepas et al. 2009), so maybe AOM activity was favored in the ambient condition. However, MBR operated in this study was inoculated by cold seep sediment from Ginsburg mud volcano obtained from 910 m water depth, so the *in situ* pressure of biomass origin was 90 times of atmospheric pressure. On a positive note, the AOM rate obtained in this study is comparable to the results achieved in a hanging sponge bioreactor that was fed with high sulfate loads (30 mmol L^{-1} day^{-1}) and operated at 10°C for 2,013 days (Aoki et al., 2014).

A decoupling between the AOM and sulfate reduction was observed in the activity assays tested in batch incubations as the sulfate consumption rate was much higher than the ^{13}C DIC production rate when incubated with ^{13}C methane. Decoupling of AOM and sulfate reduction is obvious when the amount of sulfate was limited or another electron acceptor was supplied instead of sulfate (Scheller et al., 2016). However, in this study almost 15 mM of sulfate was added to the batch which provides an environment suitable for both sulfate reducers and ANME. Even so, the value of DIC could be underestimated as with the increasing pH from the AOM conversion (Eq. 6.1) can cause a large amount of DIC produced to remain dissolved. The estimation of DIC was based on the headspace ^{13}C fraction, whereas a large amount of DIC (as $CO_{2\ (aq.)}$, H_2CO_3, HCO_3^- or CO_3^{2-}) could remain as dissolved forms (Scheller et al., 2016).

On the other hand, the total ^{13}C DIC production rate was almost 10 to 30 times higher than in the case of incubations with ^{13}C carbonate and was almost in equimolar ratio with sulfate reduction and sulfide production profiles. A larger amount of ^{13}C DIC in the headspace may be due to dissolution of an added ^{13}C carbonate in these incubations in the liquid resulting in the increment in headspace ^{13}C DIC (Scheller et al., 2016). In this study, along with rapid DIC production, the total sulfate consumption and simultaneous production of sulfide from ^{13}C carbonate incubations was also comparable to the incubations with ^{13}C methane. This clearly shows that DIC was assimilated by both ANME (Kellermann et al., 2012) and sulfate reducing bacteria (Muyzer & Stams, 2008).

6.4.2 Microbial community from Ginsburg mud volcano in the MBR

This study shows the enrichment of ANME-2 and *Desulfosarcinales* in the MBR (Figure 6.5). These two clades are usually observed together and typical to many *in situ* AOM studies (Hinrichs et al., 2000; Orphan et al., 2002; Vigneron et al., 2013) as well as AOM enrichments (Meulepas et al., 2009; Timmers et al., 2015; Zhang et al., 2011). Although none of the ANME clades were detected by Illumina Miseq in the inoculum (described in chapter 5), this study evidenced the AOM and enrichment of ANME-2 in the MBR reactor. Contrariwise, ANME-1 (40%) was mainly enriched in a biotrickling filter bioreactor inoculated with the same inoculum, with almost 10% of total archaeal community as ANME-2 (described in chapter 5). From a biotechnological perspective, the dominance of a particular ANME type after several months of operation can be influenced by the bioreactor configuration, its hydrodynamics and the physico-chemical and biological process conditions. Both the bioreactors, i.e. the biotrickling filter and the MBR, were operated in fed-batch mode of liquid mineral medium supply; however, in the MBR, the microbial biomass was in suspension, whereas biomass was attached as a biofilm onto the polyurethane sponges in the biotrickling filter. In MBR, the suspended biomass is exposed to the stress of methane bubbling and medium recirculation that could cause discrete flow patterns in the reactor (Meulepas et al. 2009).

The AOM performing microbial community in the MBR, i.e. ANME-2 and *Desulfosarcinales*, were closely associated to each other during CARD-FISH analyses which suggests a potential syntrophy between the ANME-2 and the

Desulfosarcinales. The *Desulfosarcinales* clades of sulfate reducing bacteria, which are known to be a common associate of ANME-2, were observed to be rod shaped and they could be also circular shaped such as cocci and vibrio (Schreiber et al., 2010). Nevertheless, a single shell type ANME and dispersed sulfate reducing bacteria have been observed in different ANME habitats, such as cold seeps and hydrothermal vents (Michaelis et al., 2002; Wankel et al., 2012) and ANME could potentially perform AOM alone without its sulfate reducing partner (Milucka et al., 2012; Scheller et al., 2016).

6.4.3 Role of acetate production in the MBR

In this study, intermittently, acetate production was observed in the MBR even though acetate was not fed to the MBR. The acetate production from bioreactor material was unlikely as the MBR was built constructed using a glass vessel and fitted with non-degradable Tygon tubings. The formation of acetate in a reactor or batch incubation when fed solely with methane has so far not been reported. Genetic analysis has shown that the ANME clade genome contains all the required enzymes for the reverse methanogenesis step, which also includes acetogenesis (Haroon et al., 2013; Meyerdierks et al., 2010; Wang et al., 2014). Hydrogen and other methanogenic substrates, such as acetate, formate, methanol and methanethiol have been hypothesized as the intermediates during AOM reaction (Hoehler et al., 1994; Sørensen et al., 2001; Valentine et al., 2000). However, the addition of acetate to an AOM enrichment did not enhance AOM and may favor the growth of sulfate reducers rather than ANME (Meulepas et al., 2010; Moran et al., 2008; Nauhaus et al., 2005).

The acetate formation in the MBR provides evidence on the potential involvement of acetate in the AOM process as indicated by previous hypothesis of acetate being intermediate for electron transfer between ANME and SRB (Hoehler et al., 1994; Sørensen et al., 2001; Valentine et al., 2000). Acetate was assumed to be the favorable electron shuttle in high methane pressure environments (Valentine, 2002) and even detected in high pressure incubations of the ANME-2a clades (Cassidy et al., unpublished data). Therefore, ANME may metabolize methane to acetate, which is subsequently utilized by sulfate reducing bacteria or ANME themselves and finally converted to DIC with simultaneous sulfide production. Further investigations with ^{13}C labeled acetate and ^{13}C labeled methane along with the observation of ANME cells metabolism by NanoSIMS may be performed with the MBR biomass to identify the metabolic pathways involved in acetate formation and consumption. Since the consortia were compact and seemed to be attached to each other, there could be other potential modes of electron transfer, such as direct cell to cell electron transfer (McGlynn et al., 2015). The electron transfer mechanism among ANME-2 and *Desulfosarcinales* from the MBR biomass needs to be further explored.

6.5 Conclusions

This study has shown the AOM activity and enrichment of ANME originating from cold seep environment, i.e. Ginsburg mud volcano, Gulf of Cadiz in a MBR with externally positioned ultra filtration membrane operated at ambient conditions. The enrichment of ANME-2 and *Desulfosarcinales* consortia was attained in the MBR. Previous study in MBR inoculated with shallow estuarine sediment (~ 3 atm), has indicated the high rate of AOM activity after long term operation of bioreactor. Nevertheless, this study has widened the prospective for ANME cultivation with the possibility of ANME cultivation at ambient pressure, albeit the *in situ* pressure of inoculum origin was 90 times of atmospheric pressure. So, a long term operation of MBR by inoculating diverse AOM active sediment from a wide range of marine location is suggested to achieve highly enrich ANME communities. Further, the intermittent detection of acetate suggests the acetate being formed by metabolism of methane, i.e. only carbon source added to MBR. So the acetate could be formed as an intermediate between ANME and sulfate reduction which can be further explored.

6.6 References

Acree, T.E., Sonoff, E.P., Splittstoesser, D.F. 1971. Determination of hydrogen sulfide in fermentation broths containing SO_2. *Appl. Microbiol.*, **22**(1), 110-112.

Aoki, M., Ehara, M., Saito, Y., Yoshioka, H., Miyazaki, M., Saito, Y., Miyashita, A., Kawakami, S., Yamaguchi, T., Ohashi, A., Nunoura, T., Takai, K., Imachi, H. 2014. A long-term cultivation of an anaerobic methane-oxidizing microbial community from deep-sea methane-seep sediment using a continuous-flow bioreactor. *PLoS One*, 9(8), Pe105356.

Bhattarai, S., Cassarini, C., Gonzalez-Gil, G., Egger, M., Slomp, C.P., Zhang, Y., Esposito, G., Lens, P.N.L. 2017. Anaerobic methane-oxidizing microbial community in a coastal marine sediment: anaerobic methanotrophy dominated by ANME-3. *Microb. Ecol.*, **74**(3), 608-622.

Boetius, A., Ravenschlag, K., Schubert, C.J., Rickert, D., Widdel, F., Gieseke, A., Amann, R., Jørgensen, B.B., Witte, U., Pfannkuche, O. 2000. A marine microbial consortium apparently mediating anaerobic oxidation of methane. *Nature*, **407**(6804), 623-626.

Boetius, A., Wenzhöfer, F. 2013. Seafloor oxygen consumption fuelled by methane from cold seeps. *Nat. Geosci.*, **6**(9), 725-734.

Daims, H., Brühl, A., Amann, R., Schleifer, K.-H., Wagner, M. 1999. The domain-specific probe EUB338 is insufficient for the detection of all *Bacteria*: development and evaluation of a more comprehensive probe set. *Sys. Appl. Microbiol.*, **22**(3), 434-444.

Deusner, C., Meyer, V., Ferdelman, T. 2009. High-pressure systems for gas-phase free continuous incubation of enriched marine microbial communities performing anaerobic oxidation of methane. *Biotechnol. Bioeng.*, **105**(3), 524-533.

Girguis, P.R., Cozen, A.E., DeLong, E.F. 2005. Growth and population dynamics of anaerobic methane-oxidizing archaea and sulfate-reducing bacteria in a continuous-flow bioreactor. *Appl. Environ. Microbiol.*, **71**(7), 3725-3733.

Haroon, M.F., Hu, S., Shi, Y., Imelfort, M., Keller, J., Hugenholtz, P., Yuan, Z., Tyson, G.W. 2013. Anaerobic oxidation of methane coupled to nitrate reduction in a novel archaeal lineage. *Nature*, **500**(7464), 567-570.

Hinrichs, K.-U., Summons, R.E., Orphan, V.J., Sylva, S.P., Hayes, J.M. 2000. Molecular and isotopic analysis of anaerobic methane-oxidizing communities in marine sediments. *Org. Geochem.*, **31**(12), 1685-1701.

Hoehler, T.M., Alperin, M.J., Albert, D.B., Martens, S. 1994. Field and laboratory studies of methane oxidation in an anoxic marine sediment: Evidence for a methanogen-sulfate reducer consortium. *Global Biogeochem. Cy.*, **8**(4), 451-463.

Holler, T., Widdel, F., Knittel, K., Amann, R., Kellermann, M.Y., Hinrichs, K.-U., Teske, A., Boetius, A., Wegener, G. 2011. Thermophilic anaerobic oxidation of methane by marine microbial consortia. *ISME J.*, **5**(12), 1946-1956.

Kellermann, M.Y., Wegener, G., Elvert, M., Yoshinaga, M.Y., Lin, Y.-S., Holler, T., Mollar, X.P., Knittel, K., Hinrichs, K.-U. 2012. Autotrophy as a predominant mode of carbon fixation in anaerobic methane-oxidizing microbial communities. *Proc. Nat. Acad. Sci. USA*, **109**(47), 19321-19326.

Knittel, K., Boetius, A. 2009. Anaerobic oxidation of methane: progress with an unknown process. *Annu. Rev. Microbiol.*, **63**(1), 311-334.

Knittel, K., Lösekann, T., Boetius, A., Kort, R., Amann, R. 2005. Diversity and distribution of methanotrophic archaea at cold seeps. *Appl. Environ. Microbiol.*, **71**(1), 467-479.

Krüger, M., Blumenberg, M., Kasten, S., Wieland, A., Känel, L., Klock, J.-H., Michaelis, W., Seifert, R. 2008. A novel, multi-layered methanotrophic microbial mat system growing on the sediment of the Black Sea. *Environ. Microbiol.*, **10**(8), 1934-1947.

McGlynn, S.E., Chadwick, G.L., Kempes, C.P., Orphan, V.J. 2015. Single cell activity reveals direct electron transfer in methanotrophic consortia. *Nature*, **526**(7574), 531-535.

Meulepas, R.J.W., Jagersma, C.G., Khadem, A., Stams, A.J.W., Lens, P.N.L. 2010. Effect of methanogenic substrates on anaerobic oxidation of methane and sulfate reduction by an anaerobic methanotrophic enrichment. *Appl. Microbiol. Biotechnol.*, **87**(4), 1499-1506.

Meulepas, R.J.W., Jagersma, C.G., Gieteling, J., Buisman, C.J.N., Stams, A.J.M., Lens, P.N.L. 2009. Enrichment of anaerobic methanotrophs in sulfate-reducing membrane bioreactors. *Biotechnol. Bioeng.*, **104**(3), 458-470.

Meyerdierks, A., Kube, M., Kostadinov, I., Teeling, H., Glöckner, F.O., Reinhardt, R., Amann, R. 2010. Metagenome and mRNA expression analyses of anaerobic methanotrophic archaea of the ANME-1 group. *Environ. Microbiol.*, **12**(2), 422-439.

Michaelis, W., Seifert, R., Nauhaus, K., Treude, T., Thiel, V., Blumenberg, M., Knittel, K., Gieseke, A., Peterknecht, K., Pape, T., Boetius, A., Amann, R., Jørgensen, B.B., Widdel, F., Peckmann, J., Pimenov, N.V., Gulin, M.B. 2002. Microbial reefs in the Black Sea fueled by anaerobic oxidation of methane. *Science*, **297**(5583), 1013-1015.

Milucka, J., Ferdelman, T.G., Polerecky, L., Franzke, D., Wegener, G., Schmid, M., Lieberwirth, I., Wagner, M., Widdel, F., Kuypers, M.M.M. 2012. Zero-valent sulphur is a key intermediate in marine methane oxidation. *Nature*, **491**(7425), 541-546.

Moran, J.J., Beal, E.J., Vrentas, J.M., Orphan, V.J., Freeman, K.H., House, C.H. 2008. Methyl sulfides as intermediates in the anaerobic oxidation of methane. *Environ. Microbiol.*, **10**(1), 162-173.

Muyzer, G., Stams, A.J. 2008. The ecology and biotechnology of sulphate-reducing bacteria. *Nat. Rev. Microbiol.*, **6**(6), 441-454.

Nauhaus, K., Albrecht, M., Elvert, M., Boetius, A., Widdel, F. 2007. *In vitro* cell growth of marine archaeal-bacterial consortia during anaerobic oxidation of methane with sulfate. *Environ. Microbiol.*, **9**(1), 187-196.

Nauhaus, K., Boetius, A., Kruger, M., Widdel, F. 2002. *In vitro* demonstration of anaerobic oxidation of methane coupled to sulphate reduction in sediment from a marine gas hydrate area. *Environ. Microbiol.*, **4**(5), 296-305.

Nauhaus, K., Treude, T., Boetius, A., Kruger, M. 2005. Environmental regulation of the anaerobic oxidation of methane: a comparison of ANME-I and ANME-II communities. *Environ. Microbiol.*, 7(1), 98-106.

Niemann, H., Duarte, J., Hensen, C., Omoregie, E., Magalhaes, V.H., Elvert, M., Pinheiro, L.M., Kopf, A., Boetius, A. 2006a. Microbial methane turnover at mud volcanoes of the Gulf of Cadiz. *Geochim. Cosmochim. Ac.*, **70**(21), 5336-5355.

Niemann, H., Losekann, T., de Beer, D., Elvert, M., Nadalig, T., Knittel, K., Amann, R., Sauter, E.J., Schluter, M., Klages, M., Foucher, J.P., Boetius, A. 2006b. Novel microbial communities of the Haakon Mosby mud volcano and their role as a methane sink. *Nature*, **443**, 854-858.

Orphan, V.J., House, C.H., Hinrichs, K.U., McKeegan, K.D., DeLong, E.F. 2001. Methane-consuming archaea revealed by directly coupled isotopic and phylogenetic analysis. *Science*, **293**(5529), 484-487.

Orphan, V.J., House, C.H., Hinrichs, K.-U., McKeegan, K.D., DeLong, E.F. 2002. Multiple archaeal groups mediate methane oxidation in anoxic cold seep sediments. *Proc. Nat. Acad. Sci. USA*, **99**(11), 7663-7668.

Pernthaler, A., Pernthaler, J., Amann, R. 2002. Fluorescence *in situ* hybridization and catalyzed reporter deposition for the identification of marine bacteria. *Appl. Environ. Microbiol.*, **68**(6), 3094-3101.

Pernthaler, A., Pernthaler, J., and Amann, R. 2004. Sensitive multi-color fluorescence *in situ* hybridization for the identification of environmental microorganisms. in: Kowalchuk, G., de Bruijn, F., Head, I.M., Akkermans, A.D., van Elsas, J.D. (Eds.), *Molecular Microbial Ecology Manual.* Vol. 1 and 2 (2nd Edition), Springer Netherlands, Dodrecht, the Netherlands, pp. 711-725.

Reeburgh, W.S. 2007. Oceanic methane biogeochemistry. *Chem. Rev.*, **107**(2), 486-513.

Ruff, S.E., Arnds, J., Knittel, K., Amann, R., Wegener, G., Ramette, A., Boetius, A. 2013. Microbial communities of deep-sea methane seeps at Hikurangi continental margin (New Zealand). *PLoS One*, **8**(9), e72627.

Ruff, S.E., Kuhfuss, H., Wegener, G., Lott, C., Ramette, A., Wiedling, J., Knittel, K., Weber, M. 2016. Methane seep in shallow-water permeable sediment harbors high diversity of anaerobic methanotrophic communities, Elba, Italy. *Front. Microbiol.*, **7**(374), 1-20.

Scheller, S., Yu, H., Chadwick, G.L., McGlynn, S.E., Orphan, V.J. 2016. Artificial electron acceptors decouple archaeal methane oxidation from sulfate reduction. *Science*, **351**(6274), 703-707.

Schreiber, L., Holler, T., Knittel, K., Meyerdierks, A., Amann, R. 2010. Identification of the dominant sulfate-reducing bacterial partner of anaerobic methanotrophs of the ANME-2 clade. *Environ. Microbiol.*, **12**(8), 2327-2340.

Shi, Y., Hu, S., Lou, J., Lu, P., Keller, J., Yuan, Z. 2013. Nitrogen removal from wastewater by coupling anammox and methane-dependent denitrification in a membrane biofilm reactor. *Environ. Sci. Technol.*, **47**(20), 11577-11583.

Sørensen, K., Finster, K., Ramsing, N. 2001. Thermodynamic and kinetic requirements in anaerobic methane oxidizing consortia exclude hydrogen, acetate, and methanol as possible electron shuttles. *Microb. Ecol.*, **42**(1), 1-10.

Stahl DA, Amann RI. 1991. Development and application of nucleic acid probes. in: Stackebrandt, E., Goodfellow, M. (Eds.), *Nucleic acid techniques in bacterial systematics*. John Wiley & Sons Ltd, Chichester, UK, pp 205-248.

Thauer, R.K., Shima, S. 2008. Methane as fuel for anaerobic microorganisms. *Ann. N. Y. Acad. Sci.*, **1125**(1), 158-170.

Timmers, P.H., Gieteling, J., Widjaja-Greefkes, H.A., Plugge, C.M., Stams, A.J., Lens, P.N.L, Meulepas, R.J. 2015. Growth of anaerobic methane-oxidizing archaea and sulfate-reducing bacteria in a high-pressure membrane capsule bioreactor. *Appl. Environ. Microbiol.*, **81**(4), 1286-1296

Treude, T., Boetius, A., Knittel, K., Wallmann, K., Jorgensen, B.B. 2003. Anaerobic oxidation of methane above gas hydrates at Hydrate Ridge, NE Pacific Ocean. *Mar. Ecol. Prog. Ser.*, **264**, 1-14.

Treude, T., Knittel, K., Blumenberg, M., Seifert, R., Boetius, A. 2005a. Subsurface microbial methanotrophic mats in the Black Sea. *Appl. Environ. Microbiol.*, **71**, 6375-6378.

Treude, T., Krause, S., Maltby, J., Dale, A.W., Coffin, R., Hamdan, L.J. 2014. Sulfate reduction and methane oxidation activity below the sulfate-methane transition zone in Alaskan Beaufort Sea continental margin sediments: Implications for deep sulfur cycling. *Geochim. Cosmochim. Ac.*, **144**, 217-237.

Treude, T., Krüger, M., Boetius, A., Jørgensen, B.B. 2005b. Environmental control on anaerobic oxidation of methane in the gassy sediments of Eckernförde Bay (German Baltic). *Limnol. Oceanogr.*, **50**(6), 1771-1786.

Treude, T., Orphan, V., Knittel, K., Gieseke, A., House, C.H., Boetius, A. 2007. Consumption of methane and CO_2 by methanotrophic microbial mats from gas seeps of the anoxic Black Sea. *Appl. Environ. Microbiol.*, **73**(7), 2271-2283.

Valentine, D.L. 2002. Biogeochemistry and microbial ecology of methane oxidation in anoxic environments: a review. *Anton. Leeuwen. Int. J. G.*, **81**(1-4), 271-282.

Valentine, D.L., Reeburgh, W.S., Hall, R. 2000. New perspectives on anaerobic methane oxidation. *Environ. Microbiol.*, **2**(5), 477-484.

Vigneron, A., Cruaud, P., Pignet, P., Caprais, J.-C., Cambon-Bonavita, M.-A., Godfroy, A., Toffin, L. 2013. Archaeal and anaerobic methane oxidizer communities in the Sonora Margin cold seeps, Guaymas Basin (Gulf of California). *ISME J.*, **7**(8), 1595-1608.

Vigneron, A., L'Haridon, S., Godfroy, A., Roussel, E.G., Cragg, B.A., Parkes, R.J., Toffin, L. 2015. Evidence of active methanogen communities in shallow sediments of the Sonora Margin cold seeps. *Appl. Environ. Microbiol.*, **81**(10), 3451-3459.

Villa-Gomez D, Ababneh H, Papirio S, Rousseau D.P.L., Lens P.N.L. 2011. Effect of sulfide concentration on the location of the metal precipitates in inversed fluidized bed reactors. *J. Hazard. Mater.*, **192**(1), 200-207.

Wang, F.-P., Zhang, Y., Chen, Y., He, Y., Qi, J., Hinrichs, K.-U., Zhang, X.-X., Xiao, X., Boon, N. 2014. Methanotrophic archaea possessing diverging methane-oxidizing and electron-transporting pathways. *ISME J.*, **8**(5), 1069-1078.

Wankel, S.D., Adams, M.M., Johnston, D.T., Hansel, C.M., Joye, S.B., Girguis, P.R. 2012. Anaerobic methane oxidation in metalliferous hydrothermal sediments: influence on carbon flux and decoupling from sulfate reduction. *Environ. Microbiol.*, **14**(10), 2726-2740.

Yamamoto, S., Alcauskas, J.B., Crozier, T.E. 1976. Solubility of methane in distilled water and seawater. *J. Chem. Eng. Data*, **21**(1), 78-80.

Zhang, Y., Henriet, J.-P., Bursens, J., Boon, N. 2010. Stimulation of *in vitro* anaerobic oxidation of methane rate in a continuous high-pressure bioreactor. *Biores. Technol.*, **101**(9), 3132-3138.

Zhang, Y., Maignien, L., Zhao, X., Wang, F., Boon, N. 2011. Enrichment of a microbial community performing anaerobic oxidation of methane in a continuous high-pressure bioreactor. *BMC Microbiol.*, **11**(137), 1-8.

CHAPTER 7

Response of Highly Enriched ANME-2a Community to the Different Pressure and Temperature Conditions

Abstract

Activity and growth of anaerobic methanotrophs (ANME) largely depends on the methane availability. High pressure incubation has shown to be more effective way for the study of anaerobic oxidation of methane (AOM) and enhance the growth rate of ANME. The main aim of this study is to investigate the effect of the methane partial pressure and temperature on anaerobic oxidation of methane and sulfate reduction activities by highly enriched ANME-2a community. The biomass containing ANME-2 was incubated in different pressure and temperature gradients for 80 days, i.e. 2 MPa, 10 MPa, 20 MPa and 30 MPa at 15°C to explore the AOM activity and monitor the response of microbial community at different pressures. The temperature used for the incubations were 4°C, 15°C and 25°C to study the response of the microbial community with temperature. The incubation at 10 MPa pressure and 15°C temperature showed the most active condition for the studied ANME-2 phylotype, whereas activity with 2 MPa pressure at 15°C was almost comparative to the response at 10 MPa pressure. The incubations at 20 MPa and 30 MPa pressure showed the depletion in activities after 30 days. Incubations at 4°C and 25°C at 10 MPa showed the minimum activity. Further, the microbial community analysis showed the shift in bacterial community composition after incubation in different conditions; however the archaeal community was remained stable. Thus, the finding of 15°C at 10 MPa pressure as most favorable condition for studied ANME-2 activity, suggesting the link with the biomass origin *in situ* pressure and temperature which is almost 10 MPa and 15°C respectively. The retardation of microbial activity at higher pressure of 20 MPa and 30 MPa in this study, also suggests the studied ANME cells might be influenced by increasing pressure and the ANME might not be piezophiles.

7.1 Introduction

The anaerobic oxidation of methane (AOM) coupled to sulfate reduction (AOM-SR) takes place in marine locations where sulfate plays a key role on maintaining oceanic carbon (Martens & Berner, 1974; Reeburgh, 1976). This unique microbiological phenomenon, AOM-SR is performed by anaerobic methanotrophs (ANME) and sulfate reducing bacteria belonging to *Deltaproteobacteria*. Phylogentically ANME is affiliated with various groups of methanogenic archaea and has been grouped into three distinct clades, i.e. ANME-1, ANME-2 and ANME-3 (Boetius et al., 2000; Hinrichs et al., 1999; Knittel et al., 2005; Niemann et al., 2006).

ANME are widely distributed in marine habitats including cold seep systems (Boetius & Wenzhöfer, 2013; Marlow et al., 2014; Ruff et al., 2015), hydrothermal vents (Vigneron et al., 2013; Wankel et al., 2012; Wegener et al., 2016) and organic rich sediments with diffusive methane formed by methanogenesis (Egger et al., 2015; Rooze et al., 2016; Treude et al., 2014; Treude et al., 2005). Although major cause of ANME growth and distribution has not yet been clearly understood, with the ample amount of sulfate in seawater, methane which is barely soluble in water at standard temperature and pressure could be the major restricting factor for it. Therefore, the bioreactors with elevated methane partial pressure are in use recently, to achieve higher AOM-SR rates and ANME enrichments (Deusner et al., 2009; Zhang et al., 2010; Zhang et al., 2011).

AOM depended sulfate reduction rate was increased by almost 6 times when the methane partial pressure increased from 0.2 to 6 MPa (Deusner et al., 2009). Similarly, the sulfate reduction rate of sediment from Hydrate Ridge was significantly higher at elevated methane partial pressure (Kruger et al., 2008; Nauhaus et al., 2002). ANME-2 population growth was observed when Eckernförde Bay sediment was incubated in high pressure capsules at 10 MPa (Timmers et al., 2015). All the previous studies has shown, the increment in ANME growth rate due to increment in negative gibbs free energy at higher methane partial pressure. The affinity constant (Km) for methane of ANME from Gulf of Cádiz sediment i.e. the same sediment used in this study was 37 mM which is equivalent to the total dissolved methane at 30 MPa methane partial pressure (Zhang et al., 2010). Therefore, it is advantageous to find the optimum condition for ANME growth and activity then the optimum can be applied for the future enrichments.

ANME enrichment with high methane pressure have been performed with the sediment originating from a Captain Aryutinov mud volcano from the Gulf of Cadiz, and depicted to perform AOM-SR at higher rate and yielded highly enriched ANME-2a community (Zhang et al., 2010; Zhang et al., 2011). A complete genome and functional pathway of the same enriched ANME-2a was published in recent past and all genes required for seven steps of methanogenesis from carbon dioxide were detected in ANME-2a metabolism (Wang et al., 2014). AOM and sulfate reduction rates were assessed with different co-substrates by the ANME-2a community at 100 bar incubation (Cassidy et al., unpublished data). However, the response of different conditions especially temperature and pressure to the ANME activity has not yet been well explained. Thus, the main aim of this study is to investigate the effect of the methane partial pressure and temperature on AOM and sulfate reducing activity in highly enriched ANME-2a community. Further the microbial community composition and relative abundance was assessed and compared among the different pressures and temperature regime.

7.2   Materials and methods

7.2.1 Source of biomass

Sediment was obtained from the Captain Aryutinov mud volcano (35° 39.700'N; 07° 20.012'W), Gulf of Cadiz in the Atlantic Ocean, at a water depth of 1200 m on April 2006 (Zhang et al., 2010). Prior to this activity test, the sediment biomass was enriched for ANME-2a in continuous high pressure reactor at 15°C for past four years then stored at the same condition in a high pressure vessel (Zhang et al., 2010).

7.2.2 Incubation and experimental design

The experiment was conducted in high pressure incubation bioreactors (HPB). The HPB had a capacity of 200 ml volume and could handle up to 60 MPa pressure, equipped with manometer with reading up to 100 MPa, inlets, outlets and sampling port.

The enriched slurry was homogenized and then 50 ml of the slurry was transferred into HPB. All the inoculations were performed in the anaerobic conditions under nitrogen or methane environment. Then, the incubation vessels were flushed with methane for 2 min. Subsequently, basal medium mixed with methane in a condensing vessel was added using a high pressure pump (HPLC pump) in order to achieve respective pressure. The incubation was performed at 15°C in the dark for experiment of different pressure and additional incubations were performed with 4°C and 25°C for the experiment of different temperature. The basal medium was prepared according to Zhang et al. (2010) and contained the following compounds per liter of demineralized water: NaCl (26 g), KCl (0.5 g) MgCl$_2$.6H$_2$O (5 g), NH$_4$Cl (0.3 g), CaCl$_2$.2H$_2$O (1.4 g), Na$_2$SO$_4$ (1.43 g), KH$_2$PO$_4$ (0.1 g), trace element solution (1 ml), 1 M NaHCO$_3$ (30 ml), vitamin solution (1 ml), thiamin solution (1 ml), vitamin B$_{12}$ solution (1 ml), 0.5 g L^{-1} resazurin solution (1 ml) and 0.5 M Na$_2$S solution (1 ml). Vitamins and trace element mixture was prepared according to Widdel and Bak (1992).

Both experiments i.e. AOM-SR activity in different pressure conditions and AOM-SR in different temperature conditions were performed in triplicate. The incubations for the experiment with different pressure conditions include the incubations at 2 MPa, 10 MPa, 20 MPa and 30 MPa incubations at 15°C. In addition, two sets of triplicate containers at 10 MPa were incubated at 4°C and 25°C, respectively, to assess the temperature effect. Control incubations were done at 10 MPa pressure at 15°C without any added biomass.

7.2.3 Sampling

For the sampling, each vessel was homogenized. Approximately 6 ml sample was taken by attaching a connector and a vacuum tube to the exit port while gently opening the tap. Weight and pressure of the vacuum tube before and after the sampling were measured. Pressure in each vessel was restored simultaneously by adding fresh methane mixed basal medium using the HPLC pump in order to maintain the pressure in the vessel.

7.2.4 Chemical analysis

The pH was measured using a pH indicator paper immediately after sampling. Dissolved sulfide was measured by using the methylene blue method (Hach Lange method 8131) and a DR5000 Spectrophotometer (Hach Lange GMBH, Düsseldorf, Germany). Sulfate was measured with an ion chromatograph (Metrohm 732 IC Detector) with a column (METROSEP A SUPP 5 - 250) at a flow rate of 0.7 ml min^{-1} with an 3.2 mM Na$_2$CO$_3$ / 1.0 mM NaHCO$_3$ eluent, a temperature of 35°C, a current of 13.8 mA, an injection volume of 20 µl and a retention time of 35 min.

The headspace composition of methane and carbon dioxide was measured on a gas chromatograph (GC-14B, Shimadzu, Japan). The gas chromatograph was equipped with column: TDX-02, injection temperature: 100°C, FID temperature: 250°C, methane conversion oven: 360°C. Argon was the carrier gas at a flow rate of 1.0 ml min^{-1}. Each time the headspace gas sample was measured along with the standard gas mixture of methane and carbon dioxide and measured every sample was measured in duplicate. The standard deviation showed less than 0.5% variation in the peak area of chromatograph.

7.2.5 Calculation

Total sulfide was estimated by summing up the dissolved sulfide and the sulfide concentration in gas phase. H^+ ion was estimated on the basis of pH and the sulfide in liquid phase was calculated by using H^+ concentration, amount of sulfide measured in liquid sample and the equilibrium constants. And the sulfide in the gas form was calculated according to Henry's law. Similarly, total inorgnic carbon (TIC) was estimated by accounting the inorganic carbon both in liquid and gas and estimated by using Henry's law and carbon dioxide partial pressure.

7.2.6 Microbial community analysis

Archaeal and bacterial 16S rRNA genes variable region 4 (V4) region of the 16S rRNA gene was amplified for Illumina Miseq sequencing by using Uni519F (5'- YMG CCR CGG KAA HACC-3') and Arc806R (5'-GGACTACNSGGGTMTCTAAT-3') (Song et al., 2013) and Bac533F (5'-GCCAGCAGCCGCGGTAA-3') and Bac802R (5'-TACNVGGGTATCTAATCC-3') primer sets for *Archaea* and *Bacteria* respectively. Various unique key tags for each samples was assigned each forward primer. The amplification for *Archaea* was performed as nested PCR, in which prior to amplification for V4 region, the 16s rRNA genes was amplified with universal archaeal primer pair Arc21F (5' TTC CGG TTG ATC CYG CCG GA 3') and Arc958R (5' YCC GGC GTT GAM TCC AAT T 3') as described by DeLong (1992).

The PCR reaction mixture (50 µl) contained 2 µl of DNA template (~ 70 ng) in case of *Bacteria* or amplicon from first step PCR in case of *Archaea*, 1X PCR buffer with 1.5 mM MgCl$_2$, 0.2 mM of deoxynucleoside triphosphates (dNTPs), 2.5 mg l^{-1} of Bovine serum albumin (BSA), 0.4 mM of each primer, 1.25 U of Ex Taq DNA polymerase (all Takara products, Japan) and nuclease free water. An initial denaturation step of 5 min at 94°C was followed by 30 cycles at 94°C for 40 s, 55°C for 55 sec and 72°C for

212

40 s. The final extension step was 72°C for 10 min. However, for the nested PCR with the universal archaeal primer, the PCR amplification was performed in 20 cycles.

7.2.7 Illumina Miseq data processing

The obtained sequences with length of 260 bp from illumina miseq platform were analyzed according to He et al. (2016) and Bhattarai et al. (2017). All the analysis related to sequences such as chimera elimination, clustering and classification were performed in mothur (Schloss & Westcott, 2011). Initially, the truncated and mismatch sequences were removed by using shhh.flows command of mothur. Chimeric sequences were identified and eliminated by using the chimera.uchime. The sequence reads were classified according to the Silva taxonomy (Pruesse et al., 2007) using the classify.seqs command (He et al., 2016).

7.3   Results

7.3.1 AOM-SR activity in different pressure gradient

An increase in sulfide production with simultaneous increment in sulfate reduction was observed in all the incubations with different pressure (2 MPa, 10 MPa, 20 MPa and 30 MPa) in comparison with control (Figure 7.1). During the initial 20 days all the HPB incubations with different pressures showed the similar trend with sulfide production of almost 0.5 mmol per HPB. Then, the sulfide production and sulfate reduction retarded for 20 MPa and 30 MPa incubations and oscillated within 0.5 mmol (Figures 7.1A and 7.1C). In case of HPB with 2 MPa and 10 MPa, the cumulative sulfate consumption with simultaneous sulfide production was almost five 4-6 folds higher than other incubated pressure.

The maximum sulfide production rate was for the HPB incubation at 10 MPa pressure was estimated at 0.18 mmol of sulfide day^{-1} with simultaneous sulfate reduction rate of 0.03 mmol of sulfate day^{-1}. The HPB incubation at 2 MPa, followed the similar trend of sulfide production and sulfate reduction as in the case of 10 MPa incubation with an estimated rate of 0.08 mmol of sulfide day^{-1} and 0.035 mmol of sulfate day^{-1}, respectively. Whereas, the sulfide production and sulfate consumption rate in case of incubations with 20 MPa and 30 MPa were almost half of the sulfide production rate at 10 MPa incubations.

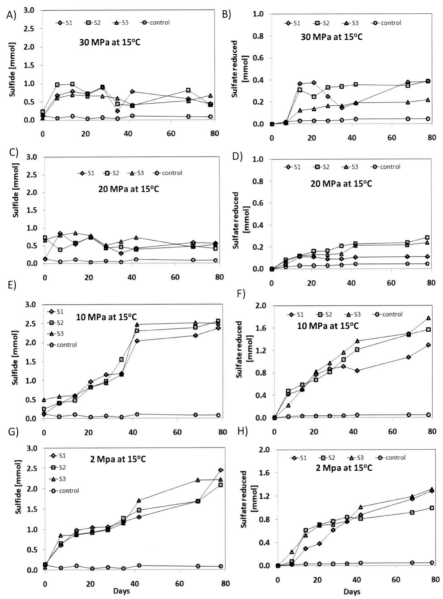

Figure 7.1 *Total accumulated sulfide and sulfate concentration in each reactor during the incubation at different pressure conditions and at 15°C A) 30 MPa, sulfide, B) 30 MPa, sulfate C) 20 MPa, sulfide, D) 20 MPa, sulfate, E) 10 MPa, sulfide, F) 10 MPa, sulfate, D) 2 MPa, sulfide and E) 2 MPa, sulfate. Here S1, S2 and S3 refers to the triplicate HPB incubations at the same conditions.*

With the comparatively lower activity among all incubations, at 20 MPa and 30 MPa, the maximum sulfide production rate was observed at the starting phase of the experiment with the value of 0.04 mmol of sulfide day^{-1} HPB^{-1} and 0.05 mmol of sulfide day^{-1} HPB^{-1}, respectively. Almost after 20 days, the sulfide production depleted and remained almost stable for both 20 MPa and 30 MPa incubations.

Similar to the sulfate reduction, the amount of methane oxidation and consecutive TIC by added methane partial pressure was higher in the case of HPB incubations at 10 MPa and 2 MPa with almost 10 mmol of cumulative TIC production in each HPB (Figures 7.3C and 7.3D). The incubations at higher pressure i.e. 20 MPa and 30 MPa showed the total TIC accumulation of almost 4 mmol during the incubation period (Figures 7.3B and 7.3D), which is exactly the half of TIC accumulated for incubations at 10 MPa and 2 MPa. Further methane in the liquid phase remained almost stable i.e. ~ 12 mmol throughout incubation period in the incubation at 20MPa and 30 MPa (Figures 7.3A and 7.3C). The control without sediment showed an almost constant TIC production of around 1 mmol of TIC, hence confirming the methane consumption was biological (Figure 7.3). The highest TIC production was estimated for incubation at 10 MPa as 0.7 mmol of TIC day^{-1} in each HPB, which suggests that the TIC accumulation rate was much higher than the sulfate reduction rate in the incubations.

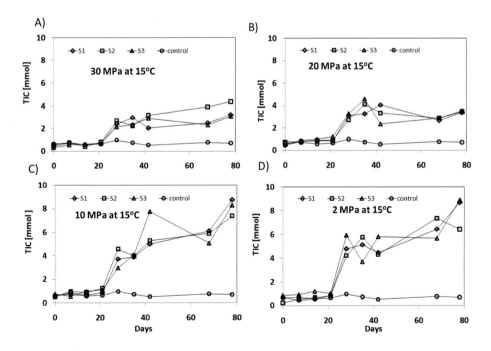

Figure 7.2 Total inorganic carbon (TIC) concentration in each reactor during the incubation at different pressure conditions and at 15°C A) 30 MPa, B) 20 MPa, C) 10 MPa, and D) 2 MPa. . Here S1, S2 and S3 refers to the triplicate HPB incubations at the same conditions.

7.3.2 AOM-SR activity in different temperature gradient

Sulfate consumption with simultaneous sulfide production in the HPB incubated at 10 MPa and 15°C was the highest among the incubations with 4°C, 15°C and 25°C temperature gradients (Figure 7.3). In both HPB incubations at 4°C and 25°C with different temperature the sulfate consumption was around 0.2 mmol up to 40 days and then it slightly increased up to 0.4 mmol (Figures 7.3B and 7.3D). Nevertheless, both sulfate consumption and sulfide production were higher than the control incubation without biomass. For 4°C HPB incubation, the maximum estimated sulfate consumption rate was 0.04 mmol of sulfate day^{-1} in each HPB with the simultaneous sulfide production rate of 0.05 mmol of sulfide day^{-1}. Similarly, at 25°C HPB incubation, the maximum sulfate reduction and sulfide reduction rate were 0.025 mmol of sulfate day^{-1} and 0.06 mmol of sulfide day^{-1}, respectively.

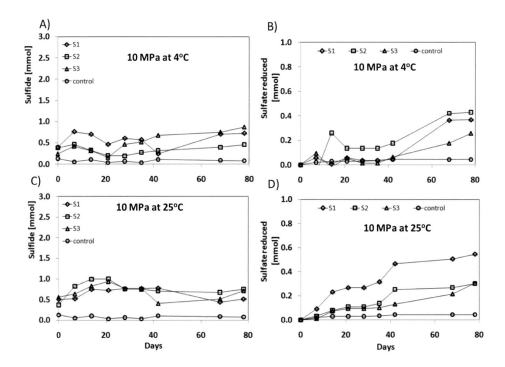

Figure 7.3 *Total accumulated sulfide and sulfate concentration in each reactor during the incubation at different temperature and at 10 MPa pressure A) 4°C, sulfide, B) 4°C, sulfate C) 25°C, sulfide and D) 25°C, sulfate. Here S1, S2 and S3 refers to the triplicate HPB incubations at the same conditions.*

TIC was in oscillating trend after 20 days of incubation for both 4°C and 25°C (Figure 7.4). Especially, for incubation at 25°C, there was an abrupt rise in the TIC production from day 21 to day 28 with an estimated rate of almost 1 mmol of TIC production day[-1]. Then again, the cumulative TIC decreased after day 28 and again increased so that the total cumulative TIC production at 25°C HPB incubation was 8.7 mmol of TIC at the end of incubation. For the incubation at 4°C, total cumulative TIC production remained below 4 mmol during the incubation period with the maximum rate of TIC production of 0.18 mmol of TIC day^{-1}, which was almost five times lower than the TIC production rate at 15°C and 25°C HPB incubations.

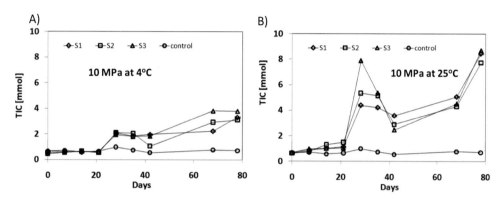

Figure 7.4 *Total inorganic carbon (TIC) concentration in each reactor during the incubation at different temperature conditions and at 10 MPa pressure A) 4°C and B) 25°C. Here S1, S2 and S3 refers to the triplicate HPB incubations at the same conditions*

7.3.3 Archaeal community in different pressure and temperature gradient

A total of 50,000 (±10,000) sequences reads were obtained for archaeal 16S rRNA gene amplicon for both in initial inoculum and the biomass after incubations at different temperature and pressure gradient via high-throughput sequencing. *Euryarchaeota* (almost 90% of relative abundance), Miscellaneous Crenarachaeotic Group and *Thaumarchaeota* were the major phylotypes in all the biomass analyzed. Among *Euryarchaeota*, the majority of sequences were affiliated to ANME-2a for which the biomass was enriched for four years prior to this study. Other euryarchaeotal clades present mostly in both biomass obtained prior to the incubation and after the incubations were; ANME clades related to ANME- 2b, Marine Group II, *Thermococcaceae* and *Methnococcoides* (Figure 7.5A). The major shift in archaeal community composition was not observed before and after the incubations in different pressure and temperature condition (Figure 7.5A).

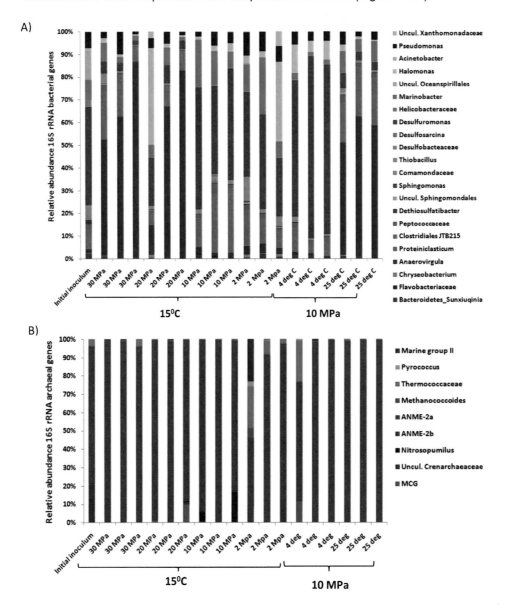

Figure 7.5 *Microbial community composition as derived by Illumina Miseq method of biomass obtained from initial inoculums and at the end of incubation in high pressure bioreactor for 80 days with different temperature and pressure A) Bacteria B) Archaea*

7.3.4 Bacterial community in different pressure and temperature gradient

Among the triplicate analysis from each condition, all showed similar archaeal community composition (Figure 7.5A). Mainly, with the comparatively higher AOM-SR activity for the incubation at 10 MPa and 15°C, the archaeal community mainly consisted of ANME-2a (85 to 99%, relative abundance) and *Nitrosopumilus* order of *Thaumarchaeota* (5-16%, relative abundance). In case of incubation at 2 MPa, other euryarchaeotal phylotypes such as; Marine Group II and *Thermococcaceae* were also found, however as similar to other biomass majority of sequences were related to ANME-2a. Further, the incubations at higher pressure condition, i.e. 20 MPa and 30 MPa, showed the highest relative abundance of almost 99% of ANME-2a. Similarly, the incubation at 25°C and 10 MPa also showed almost 100% relative abundance of ANME-2a community among 16S rRNA archaeal genes.

The bacterial high throughput sequence analysis (Figure 7.4B) showed a comparatively larger diversity than archaeal diversity with retrieval 16S rRNA bacterial genes affiliated to more than different twenty genuses (Figure 7.4B). These diverse clade of *Bacteria* were distributed amoung the major phyla of *Proteobacteria*, *Firmicutes* and *Bacteroidetes*. Among the retrieved phyla, the *Proteobacteria* showed the highest relative abundance (more than 60%) in the analyzed biomass including initial inoculum.

The most active HPB incubation in chemical analysis i.e. 10 MPa and 2 MPa showed the similar bacterial community composition with the higher relative abundance (40-55%) of *Desulfuromonas* clade of *Deltaprotoebacteria*. Similarly, *Desulfuromonas* was in majority in case of initial inoculum and in the HPB incubations at 4°C. Further, *Desulfosarcina*, a known sulfate reducing partner of ANME was also retrieved in the HPB incubations at 10 MPa and 2 MPa, however the relative abundance was less than 2%. *Desulfosarcina* were also present in initial inoculum with the relative abundance of almost 0.5%. Further, other bacterial genus i.e. *Thiobacillus* with the potential role in sulfur metabolism was found in HPB incubation at 10 MPa and 2 MPa, with the relative abundance of 1-2%. Other major bacterial clade found in HPB incubation at 10 MPa and 2 MPa were *Marinobacter* from *Gammaproteobacteria* phylum and *Clostridiales JTB215*.

The shift in bacterial community composition was observed in the incubation at higher pressure, i.e. 20 MPa and 30 MPa as *Anaerovirgula* of *Clostridiales* family were in majority. However, *Desulfuromonas* were also found in both incubations with the relative abundance less than 5% at 30 MPa and less than 10% at 20 MPa incubations. Another genus of *Clostridiales* i.e. *Clostridiales JTB215* and *Pseudomonas* were also found with the relative abundance of 5-20%, moreover, *Marinobacter* was also found in some of the incubations at 20 MPa and 30 MPa. All three biomass samples analyzed from the HPB incubations at 25°C showed the similar bacterial community composition as in the case of 30 MPa with the almost 80% relative abundance of *Clostridiales* family. Other bacterial clades present in initial innoculum such as *Marinobacter* and *Desulfuromonas* were also present, however less than 5% in relative abundance.

7.4 Discussion

7.4.1 Effect of pressure gradient in AOM-SR

This study shows that the incubation at 10 MPa and 15°C was the most active in AOM-SR, suggesting that the enriched ANME-2a community was adopted to the almost similar pressure at the *in situ* condition of biomass origin i.e. Captain Aryutinov mud volcano, which is located approximal 900 m water depth (Zhang et al., 2010). Unlike to our assumption, AOM-SR could be boosted with the increment of methane partial pressure, the AOM-SR activity slightly retarded in the HPB incubations at higher pressures, i.e. 20 MPa and 30 MPa. Majority of previous high pressure batch incubations and bioreactors based study had shown the increment in AOM-SR activities with the increase in methane partial pressure (Deusner et al., 2009; Krüger et al., 2008; Timmers et al., 2015). Nevertheless, all of these previous studies were performed at the pressures up to 10 MPa, and the data has shown the apparent affinity of methane was around 37 mM. So in this study as well, there could be the saturation of methane after 10 MPa so resulting in the active AOM up to 10 MPa and then again the activity depleted by the influence of pressure in cells. So far this is only detailed study to explore the AOM-SR activity in different pressure, pressure stress on ANME cells can be explored further.

A vast majority deep sea microbes are piezophilic or piezotolerant microorganisms from the deep-sea, including the microbial community from seep sediments (Zhang et al., 2015). However, the finding of retardation of AOM-SR activity suggest that possibly ANME were not obligate pressure loving microbes i.e. piezophilic species that cannot grow at atmospheric pressure (Kato et al., 2008; Kim & Kato, 2010). Similar to this study result, *Desulfovibrio indonesiensis* strain of sulfate reducing bacteria isolated from deep subsurface of the Juan de Fuca Ridge showed the maximum growth rate at *in situ* condition of its origin i.e. 30 MPa and temperature of around 45°C (Fichtel et al., 2015). The depletion of AOM-SR activity after 30 days of incubation at 30 MPa suggest that, the AOM enrichment used for this study is rather non-piezophilic (Fang et al., 2010), albeit the methane partial increases in two fold while increasing the pressure from 10 MPa to 30 MPa. Among different isolated bacterial and archaeal strains obtained from deep-sea environment with elevated hydrostatic pressure and harsh temperature, only 25% of them were found to be true piezophiles and theromophiles (Jebbar et al., 2015). High pressure can impact the structure of several cellular components and functions, such as membrane fluidity, protein activity and structure (Jebbar et al., 2015), however the microbial community can thrive with the minimum energy requirement to sustain in such harsh condition (Xiao & Zhang, 2014). Therefore, further study to explore the stress created by increasing pressure to ANME cells and their activity by using highly enriched ANME community or cultivated strain can decipher on the preference of ANME towards the elevated pressure.

7.4.2 Effect of temperature in AOM-SR

Likewise the pressure, the studied ANME enrichment appeared to be more adapted to the *in situ* temperature, i.e. almost 15°C. Our study result of HPB incubation at 25°C has shown almost similar effect on AOM-SR as in the case of HPB incubations with elevated pressures. The microbial metabolisms based on several chemical reactions can be influenced by each degree of rise and fall in temperature. Anaerobic methotrophy is considered as a reverse methogenesis process and methanogens are normally mesophilic and performs well even at higher temperature therefore, ANME may also perform well at slightly higher temperature than the *in situ* condition (Xiao & Zhang, 2014). However, this study with the ANME enrichment from cold seep environment showed that the activity depleted with even rise of *in situ* temperature by 10°C.

ANME has been retrieved from the environments with a wide range of temperature i.e. from -2 to 100°C in different marine locations (Holler et al., 2011; Ruff et al., 2015; Wankel et al., 2012), so the ANME could possibly adopt themselves up to certain addition or depletion of temperature. Among three major ANME clades, ANME-1 clade was found in a wide range of temperature and shown to perform even at 90°C (Holler et al., 2011; Wankel et al., 2012). Nevertheless, ANME-2 clade, were mostly retrieved from cold seep environments and shown its preference of temperature at 2 to 20°C (Marlow et al., 2014; Mason et al., 2015; Orphan et al., 2001a; Orphan et al., 2001b; Orphan et al., 2002). Although the activity and growth of each ANME type at various temperatures has not been detailed yet, ANME-2 was found to be more active at temperature lower than 20°C. Similar to our study result, bioreactor enrichments of ANME-2 with Eckernförde Bay sediment showed the maximum AOM rate when the bioreactor was operated at 15°C rather than at 30°C (Meulepas et al., 2009). Further, similar result was obtained for Eckernförde Bay *in vitro* AOM rate measurements, which showed a steady increment in AOM rates from 4°C to 20°C and subsequently decreased afterwards (Treude et al., 2005).

7.4.3 Microbial community shift with different environmental conditions

This study finding showed that the archaeal community composition remained almost stable even after the incubation at different temperature and pressure conditions for more than two months. Nevertheless, there was a distinct change in bacterial community composition with the major bacterial community shift for the HPB incubations at 30 MPa with 15°C and also for 25°C incubation at 10 MPa.

Archaea, known as the origin of life was also called as extremophiles has shown its robustness towards the harsh environmental conditions for example; high temperature and hydrostatic pressure in marine hydrothermal vents (Xiao & Zhang, 2014). However, the prokaryotic community may change over a small variation of temperature which is the direct indication of impact of temperature in the microbial metabolisms (Roussel et al., 2015). In this study, the extremely slow growing *Archaea* i.e. ANME with the average doubling time of 1.5 to 7 months (Knittel &

Boetius, 2009) was studied for almost 2.5 months under different pressure and temperature gradient, therefore, the major shift in archaeal community composition was not expected. Albeit, a clear distinction in the chemical activity was observed among the incubations at different conditions, the ANME community may take several months for its complete shift.

Further, in case of marine anaerobic *Bacteria* with the average doubling time of less than an hour to few days (Mann & Lazier, 2013), the shift on bacterial community composition was assumed during the HPB incubation period. A shift in bacterial community was observed in a well-structured pattern. Dominancy of *Desulfuromonas*, a known bacterial clade to reduce sulfur to sulfide (Greene, 2014) was observed in majority for the incubations at 10 MPa and 2 MPa, which showed the highest AOM-SR. This finding also suggest towards the formation of elemental sulfur during AOM-SR process as postulated by Millucka et al. (2012), which ultimately can be utilized by sulfur reducing bacteria such as *Desulfuromonas* in case of this study. The *Desulfuromonas* community seems to be invaded by the *Clostradiales,* with the pressure increment and also at 25°C. Therefore, our assumption was in agreement to the previous finding (Milucka et al., 2012), as shown by the chemical activity result the ANME activity was hindered with increasing pressure i.e. more than 10 MPa thus the formation of elemental sulfur as an intermediate of AOM process was decreased so the microbial metabolisms of *Desulfuromonas* community was not supported in HPB. On the other side, *Anaerovirgula* a spore forming *Bacteria* with the growth temperature range of 10-45°C and mostly growing in alkaline condition (Pikuta et al., 2006; Roof et al., 2013; Sousa et al., 2015), proliferated in the HPB incubated at higher pressures and temperature.

7.5 Conclusions

AOM activity and changes in microbial community composition of a highly enriched ANME-2a community was assessed in different temperature and pressure gradient. The highest AOM-SR activity was achieved in the incubation at 10 MPa and 15°C among the other pressure incubations (2 MPa, 20 MPa and 30 MPa). The highest activity was obtained at the condition which was similar to *in situ* of biomass origin suggesting that the ANME communities are better adapted to their natural habitat conditions. The ANME cells were more likely influenced by increment of pressure and temperature resulting in lower activity, suggesting ANME may not be piezophiles. Further, the shift in bacterial community composition after 80 days incubations at different temperature and pressure was observed. Thus, the changed microbial community composition above 20 MPa pressure at 15°C and the incubation at 25°C with 10 MPa pressure could be the indication of the change in metabolic pathway in the HPB by the influence of pressure and temperature.

7.6 References

Bhattarai, S., Cassarini, C., Gonzalez-Gil, G., Egger, M., Slomp, C.P., Zhang, Y., Esposito, G., Lens, P.N.L. 2017. Anaerobic methane-oxidizing microbial community in a coastal marine sediment: anaerobic methanotrophy dominated by ANME-3. *Microb. Ecol.*, **74**(3), 608-622.

Boetius, A., Ravenschlag, K., Schubert, C.J., Rickert, D., Widdel, F., Gieseke, A., Amann, R., Jorgensen, B.B., Witte, U., Pfannkuche, O. 2000. A marine microbial consortium apparently mediating anaerobic oxidation of methane. *Nature*, **407**(6804), 623-626.

Boetius, A., Wenzhöfer, F. 2013. Seafloor oxygen consumption fuelled by methane from cold seeps. *Nat. Geosci.*, **6**(9), 725-734.

DeLong, E.F. 1992. Archaea in coastal marine environments. *Proc. Natl. Acad. Sci. USA*, **89**(12), 5685-5689.

Deusner, C., Meyer, V., Ferdelman, T. 2009. High-pressure systems for gas-phase free continuous incubation of enriched marine microbial communities performing anaerobic oxidation of methane. *Biotechnol. Bioeng.*, **105**(3), 524-533.

Egger, M., Rasigraf, O., Sapart, C.I.J., Jilbert, T., Jetten, M.S., Röckmann, T., Van der Veen, C., Banda, N., Kartal, B., Ettwig, K.F., Slomp, C.P. 2015. Iron-mediated anaerobic oxidation of methane in brackish coastal sediments. *Environ. Sci. Technol.*, **49**(1), 277-283.

Fang, J., Zhang, L., Bazylinski, D.A. 2010. Deep-sea piezosphere and piezophiles: geomicrobiology and biogeochemistry. *Trends Microbiol.*, **18**(9), 413-422.

Fichtel, K., Logemann, J., Fichtel, J., Rullkötter, J., Cypionka, H., Engelen, B. 2015. Temperature and pressure adaptation of a sulfate reducer from the deep subsurface. *Front. Microbiol.*, **6**(1078), 1-13.

Greene, A.C. 2014. The Family Desulfuromonadaceae. in: Rosenberg E., DeLong E. F., Lory S., Stackebrandt E., Thompson F. (Eds.), *The Prokaryotes*. Springer Berlin Hiedelberg, Germany, pp. 143-155.

He, Y., Li, M., Perumal, V., Feng, X., Fang, J., Xie, J., Sievert, S., Wang, F. 2016. Genomic and enzymatic evidence for acetogenesis among multiple lineages of the archaeal phylum Bathyarchaeota widespread in marine sediments. *Nat. Microbiol.*, **1**, 16035.

Hinrichs, K.U., Hayes, J.M., Sylva, S.P., Brewer, P.G., DeLong, E.F. 1999. Methane-consuming archaebacteria in marine sediments. *Nature*, **398**(6730), 802-805.

Holler, T., Widdel, F., Knittel, K., Amann, R., Kellermann, M.Y., Hinrichs, K.-U., Teske, A., Boetius, A., Wegener, G. 2011. Thermophilic anaerobic oxidation of methane by marine microbial consortia. *ISME J.*, **5**(12), 1946-1956.

Jebbar, M., Franzetti, B., Girard, E., Oger, P. 2015. Microbial diversity and adaptation to high hydrostatic pressure in deep-sea hydrothermal vents prokaryotes. *Extremophiles*, **19**(4), 721-740.

Kato, C., Nogi, Y., Arakawa, S. 2008. Isolation, cultivation, and diversity of deep-sea piezophiles. *High-Pressure Microbiol*, **10**(1128), 203-217.

Kim, S.-J., Kato, C. 2010. Sampling, isolation, cultivation, and characterization of piezophilic microbes. in: Timmis, K. N. (Ed.), *Handbook of Hydrocarbon and Lipid Microbiology*. Springer Berlin Hiedelberg,Germany, pp. 3869-3881.

Knittel, K., Boetius, A. 2009. Anaerobic oxidation of methane: progress with an unknown process. *Annu. Rev. Microbiol.,* **63**(1), 311-334.

Knittel, K., Lösekann, T., Boetius, A., Kort, R., Amann, R. 2005. Diversity and distribution of methanotrophic archaea at cold seeps. *Appl. Environ. Microbiol.,* **71**(1), 467-479.

Krüger, M., Blumenberg, M., Kasten, S., Wieland, A., Känel, L., Klock, J.-H., Michaelis, W., Seifert, R. 2008. A novel, multi-layered methanotrophic microbial mat system growing on the sediment of the Black Sea. *Environ. Microbiol.*, **10**(8), 1934-1947.

Krüger, M., Wolters, H., Gehre, M., Joye, S.B., Richnow, H.-H. 2008. Tracing the slow growth of anaerobic methane-oxidizing communities by ^{15}N-labelling techniques. *FEMS Microbiol. Ecol.*, **63**(3), 401-411.

Mann, K.H., Lazier, J.R. 2013. *Dynamics of marine ecosystems: biological-physical interactions in the oceans.* (3rd edition), John Wiley & Sons, Malden MA and Oxford UK.

Marlow, J.J., Steele, J.A., Ziebis, W., Thurber, A.R., Levin, L.A., Orphan, V.J. 2014. Carbonate-hosted methanotrophy represents an unrecognized methane sink in the deep sea. *Nat. Commun.*, **5**(5094), 1-12.

Martens, C.S., Berner, R.A. 1974. Methane production in the interstitial waters of sulfate-depleted marine sediments. *Science*, **185**(4157), 1167-1169.

Mason, O.U., Case, D.H., Naehr, T.H., Lee, R.W., Thomas, R.B., Bailey, J.V., Orphan, V.J. 2015. Comparison of archaeal and bacterial diversity in methane seep carbonate nodules and host sediments, Eel River Basin and Hydrate Ridge, USA. *Microbial. Ecol.*, **70**(3), 766-784.

Meulepas, R.J.W., Jagersma, C.G., Gieteling, J., Buisman, C.J.N., Stams, A.J.M., Lens, P.N.L. 2009. Enrichment of anaerobic methanotrophs in sulfate-reducing membrane bioreactors. *Biotechnol. Bioeng.*, **104**(3), 458-470.

Milucka, J., Ferdelman, T.G., Polerecky, L., Franzke, D., Wegener, G., Schmid, M., Lieberwirth, I., Wagner, M., Widdel, F., Kuypers, M.M.M. 2012. Zero-valent sulphur is a key intermediate in marine methane oxidation. *Nature*, **491**(7425), 541-546.

Nauhaus, K., Boetius, A., Kruger, M., Widdel, F. 2002. *In vitro* demonstration of anaerobic oxidation of methane coupled to sulphate reduction in sediment from a marine gas hydrate area. *Environ. Microbiol.*, **4**(5), 296-305.

Niemann, H., Losekann, T., de Beer, D., Elvert, M., Nadalig, T., Knittel, K., Amann, R., Sauter, E.J., Schluter, M., Klages, M., Foucher, J.P., Boetius, A. 2006. Novel microbial communities of the Haakon Mosby mud volcano and their role as a methane sink. *Nature*, **443**(7113), 854-858.

Orphan, V.J., Hinrichs, K.-U., Ussler, W., Paull, C.K., Taylor, L.T., Sylva, S.P., Hayes, J.M., Delong, E.F. 2001a. Comparative analysis of methane-oxidizing archaea and sulfate-reducing bacteria in anoxic marine sediments. *Appl. Environ. Microbiol.*, **67**(4), 1922-1934.

Orphan, V.J., House, C.H., Hinrichs, K.U., McKeegan, K.D., DeLong, E.F. 2001b. Methane-consuming archaea revealed by directly coupled isotopic and phylogenetic analysis. *Science*, **293**(5529), 484-487.

Orphan, V.J., House, C.H., Hinrichs, K.-U., McKeegan, K.D., DeLong, E.F. 2002. Multiple archaeal groups mediate methane oxidation in anoxic cold seep sediments. *Proc. Nat. Acad. Sci. USA*, **99**(11), 7663-7668.

Pikuta, E.V., Itoh, T., Krader, P., Tang, J., Whitman, W.B., Hoover, R.B. 2006. *Anaerovirgula multivorans* gen. nov., sp. nov., a novel spore-forming, alkaliphilic anaerobe isolated from Owens Lake, California, USA. *I. J. Sys. Evol. Microbiol.* **56**(11), 2623-2629.

Pruesse, E., Quast, C., Knittel, K., Fuchs, B.M., Ludwig, W., Peplies, J., Glöckner, F.O. 2007. SILVA: a comprehensive online resource for quality checked and aligned ribosomal RNA sequence data compatible with ARB. *Nucleic Acids Res.*, **35**(21), 7188-7196.

Reeburgh, W.S. 1980. Anaerobic methane oxidation: rate depth distributions in Skan Bay sediments. *Earth Planet. Sci. Lett.*, **47**(3), 345-352.

Roof, E., Pikuta, E., Otto, C., Williams, G., Hoover, R. 2013. Some unique features of alkaliphilic anaerobes. *Proc. SPIE* 8865, Instruments, Methods, and Missions for Astrobiology XVI, 88650F (September 26, 2013), doi:10.1117/12.2045350.

Rooze, J., Egger, M., Tsandev, I., Slomp, C.P. 2016. Iron-dependent anaerobic oxidation of methane in coastal surface sediments: Potential controls and impact. *Limnol. Oceanogr.* doi: 10.1002/lno.10275.

Roussel, E.G., Cragg, B.A., Webster, G., Sass, H., Tang, X., Williams, A.S., Gorra, R., Weightman, A.J., Parkes, R.J. 2015. Complex coupled metabolic and prokaryotic community responses to increasing temperatures in anaerobic marine sediments: critical temperatures and substrate changes. *FEMS Microbiol. Ecol.*, **91**(8), doi: 10.1093/femsec/fiv084.

Ruff, S.E., Biddle, J.F., Teske, A.P., Knittel, K., Boetius, A., Ramette, A. 2015. Global dispersion and local diversification of the methane seep microbiome. *Proc. Nat. Acad. Sci. USA.*, **112**(13), 4015-4020.

Schloss, P.D., Westcott, S.L. 2011. Assessing and improving methods used in operational taxonomic unit-based approaches for 16S rRNA gene sequence analysis. *Appl. Environ. Microbiol.*, **77**(10), 3219-3226.

Song, Z.-Q., Wang, F.-P., Zhi, X.-Y., Chen, J.-Q., Zhou, E.-M., Liang, F., Xiao, X., Tang, S.-K., Jiang, H.-C., Zhang, C.L., Dong, H., Li, W.-J. 2013. Bacterial and archaeal diversities in Yunnan and Tibetan hot springs, China. *Environ. Microbiol.*, **15**(4), 1160-1175.

Sousa, J.A., Sorokin, D.Y., Bijmans, M.F., Plugge, C.M., Stams, A.J. 2015. Ecology and application of haloalkaliphilic anaerobic microbial communities. *Appl. Microbiol. Biotechnol.*, **99**(22), 9331-9336.

Timmers, P.H., Gieteling, J., Widjaja-Greefkes, H.A., Plugge, C.M., Stams, A.J., Lens, P.N., Meulepas, R.J. 2015. Growth of anaerobic methane-oxidizing archaea and sulfate-reducing bacteria in a high-pressure membrane capsule bioreactor. *Appl. Environ. Microbiol.*, **81**(4), 1286-1296.

Treude, T., Krause, S., Maltby, J., Dale, A.W., Coffin, R., Hamdan, L.J. 2014. Sulfate reduction and methane oxidation activity below the sulfate-methane transition zone in Alaskan Beaufort Sea continental margin sediments: Implications for deep sulfur cycling. *Geochim. Cosmochim. Ac.*, **144**, 217-237.

Treude, T., Krüger, M., Boetius, A., JÃ¸rgensen, B.B. 2005. Environmental control on anaerobic oxidation of methane in the gassy sediments of Eckernförde Bay (German Baltic). *Limnol.Oceanogr.*, **50**(6), 1771-1786.

Vigneron, A., Cruaud, P., Pignet, P., Caprais, J.-C., Cambon-Bonavita, M.-A., Godfroy, A., Toffin, L. 2013. Archaeal and anaerobic methane oxidizer communities in the Sonora Margin cold seeps, Guaymas Basin (Gulf of California). *ISME J.*, **7**(8), 1595-1608.

Wang, F.-P., Zhang, Y., Chen, Y., He, Y., Qi, J., Hinrichs, K.-U., Zhang, X.-X., Xiao, X., Boon, N. 2014. Methanotrophic archaea possessing diverging methane-oxidizing and electron-transporting pathways. *ISME J.*, **8**(5), 1069-1078.

Wankel, S.D., Adams, M.M., Johnston, D.T., Hansel, C.M., Joye, S.B., Girguis, P.R. 2012. Anaerobic methane oxidation in metalliferous hydrothermal sediments: influence on carbon flux and decoupling from sulfate reduction. *Environ. Microbiol.*, **14**(10), 2726-2740.

Wegener, G., Krukenberg, V., Ruff, S.E., Kellermann, M.Y., Knittel, K. 2016. Metabolic capabilities of microorganisms involved in and associated with the anaerobic oxidation of methane. *Front. Microbiol.*, **7**(46), 1-16.

Widdel, F., Bak, F. 1992. Gram negative mesophilic sulfate reducing bacteria. in: Balows, A., Truper, H., Dworkin, M., Harder, W., Schleifer, K.H. (Eds.), *The prokaryotes: a handbook on the biology of bacteria: ecophysiology, isolation, identification, applications,*Vol. II. Springer New York, USA, pp. 3352-3378.

Xiao, X., Zhang, Y. 2014. Life in extreme environments: Approaches to study life-environment co-evolutionary strategies. *Sci. China Earth Sci.*, **57**(5), 869-877.

Zhang, Y., Henriet, J.-P., Bursens, J., Boon, N. 2010. Stimulation of in vitro anaerobic oxidation of methane rate in a continuous high-pressure bioreactor. *Biores. Technol.*, **101**(9), 3132-3138.

Zhang, Y., Li, X., Bartlett, D.H., Xiao, X. 2015. Current developments in marine microbiology: high-pressure biotechnology and the genetic engineering of piezophiles. *Curr. Opin. Biotechnol.*, **33**, 157-164.

Zhang, Y., Maignien, L., Zhao, X., Wang, F., Boon, N. 2011. Enrichment of a microbial community performing anaerobic oxidation of methane in a continuous high-pressure bioreactor. *BMC Microbiol.* **11**(1), 1-8.

CHAPTER 8

General Discussion

8.1   Introduction

Anaerobic oxidation of methane (AOM) coupled to the reduction of sulfate in marine sediments is a microbe-mediated biogeochemical process. This process prevents the emission of methane to the atmosphere and has potential applications in biotechnology. In-depth knowledge on the microorganisms responsible for the process and the metabolic pathway remain, however, poorly understood. The different thesis chapters describe the activity and diversity of anaerobic methanotrophs (ANME) from marine sediments and cold seeps and their enrichment in different bioreactor configurations. Despite adopting robust enrichment procedures, the slow growth rate of ANME (1-7 months), limited solubility of methane and difficulty of handling the deep sea sediments in the laboratory remain major strides for ANME cultivation.

Previous AOM studies performed under *in situ* and *ex situ* conditions have been overviewed in Chapter 2. Although, it has been almost 50 years since AOM was discovered, bioreactor based ANME enrichment studies are still in early stage of development. Most of the bioreactor based studies were performed only in the recent decades. The chapters presented in this dissertation are an in-depth analysis of activity, microbial diversity and ANME enrichment in different bioreactor configuration. The major findings of this thesis are schematically summarized in Figure 8.1.

8.2   Anaerobic oxidation of methane in coastal sediment

In the past decades AOM research was primarily focused on the exploration of possible AOM occurrence sites and characterizing ANME types and their closer associates of sulfate reducing bacteria (SRB) in deep sea locations (Boetius et al., 2000; Sivan et al., 2007) and some extreme locations like cold seeps (Boetius & Wenzhofer, 2013; Orphan et al., 2004; Ruff et al., 2016), mud volcanoes (Niemann et al., 2006) and hydrothermal vents (Biddle et al., 2011; Brazelton et al., 2006). Study of deep sea has its own limitation such as locating the site which requires highly sophisticated submersible technologies (Ussler III et al., 2013), coring and core handling from the sea bottom to the laboratory by maintaining anaerobic conditions

(Thornburg et al., 2010; Thurber et al., 2014) and enrichment by mimicking the conditions prevailing in the sea bed or sediment layers (Boetius et al., 2009; Meulepas et al., 2010; Zhang et al., 2011a).

These constraints of mimicking *in situ* temperatures, pressures and AOM activities from the deep sea to the laboratory have hindered the ANME study and make the enrichment or culture of ANME more challenging. Therefore, the exploration of shallow coastal sediments (shallower than 100 m water depth) in terms of its capability to perform AOM can be advantageous for obtaining the AOM hosting sediments, understanding their AOM and sulfate reduction rates and enriching them in the laboratory.

In chapters 3 and 4 of this thesis, we investigated the capability of shallow coastal sediment from Marine Lake Grevelingen (the Netherlands) to perform AOM and sulfate reduction. AOM and sulfate reducing capability of the coastal sediment from the Marine Lake Grevelingen (the Netherlands) using different electron donors and electron acceptors was demonstrated in Chapters 3 and 4. The microbial activity tests have shown the consumption of methane as well as usage of a wide range of electron donors by the sediment microbiota. Therefore, detailed microbial community characterization was performed to check the potential microbial clade playing a role in carbon and sulfur metabolism (Chapter 4).

In contrast to marine sediments, limited information is available on the microbial communities involved in AOM coupled to sulfate reduction (AOM-SR) in coastal subsurface sediments. Although such environments might offer great biodiversity and may harbor different microbial communities, there is particular concern among the research community because of the frequent detection of non-cultivable microorganisms and on the reliability of microorganisms that have been cultured, identified and reported in the literature (Dale et al., 2008; Oni et al., 2015; Treude et al., 2005).

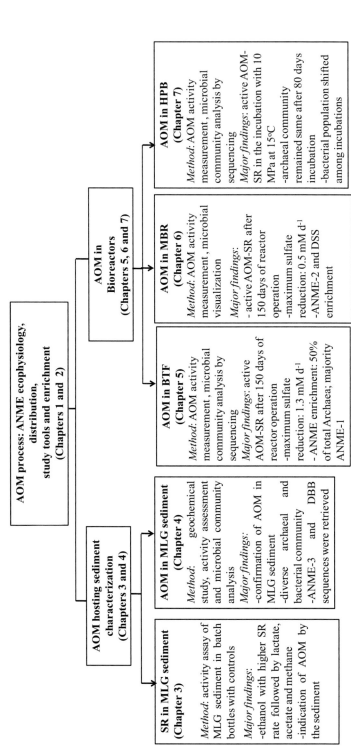

Figure 8.1 *Summary of the main methods and important observations of this thesis related to characterization of anaerobic methane oxidation hosting coastal sediments and the enrichment of anaerobic methanotrophs in bioreactors. Note: AOM - anaerobic oxidation of methane, SR - sulfate reduction, MLG - Marine Lake Grevelingen, BTF - biotrickling filter, MBR - membrane bioreactor, HPB- High pressure bioreactor, DSS - Desulfosarcinales and DBB - Desulfobulbus.*

Except the retrieval of few ANME-3 clones from sediments from the Helgoland mud area (North Sea) (Oni et al., 2015), almost all shallow subsurface sediments have shown the prevalence of ANME-2 and *Desulfosarcina* consortia as the key microbial community responsible for performing the AOM (Aquilina et al., 2010; Dale et al., 2008; Treude et al., 2005).In Chapter 4, on the contrary, we found the ANME-3 archaea and the *Desulfobulbus* clade of SRB to perform AOM in coastal sediment from Marine Lake Grevelingen (Bhattarai et al., 2017). As reported previously, these clades of ANME were only observed in some specific mud volcanoes and cold seep, such as, Haakon Mosby mud volcano and Eastern Mediterranean seepages (Heijs et al., 2007; Losekann et al., 2007; Niemann et al., 2006; Pachiadaki et al., 2010). We demonstrated that the ANME and SRB from Marine Lake Grevelingen sediment appeared as a separate cluster, thus providing strong evidence for the possible involvement of ANME-3 solely in AOM, rather than the syntrophy or the association of two microbial partners living in close metabolic interdependence as discussed in Chapter 4.

The determination of ANME-3 in Chapter 4 has widened the spatial distribution pattern of ANME-3 in diverse marine locations including coastal sediments. Retrieval of ANME-3 in our study from the shallow anoxic surface sediment (~ 45 m of water depth) with high sedimentation rates has shown the potentiality of enriching them under standard atmospheric pressure or nearly 4.5 times the atmospheric pressure under laboratory conditions. Researchers have also used high pressure bioreactors to simulate field conditions, i.e. deep sea sediment conditions. These are usually effective for enriching ANME and achieving AOM-SR activities (Deusner et al., 2009; Nauhaus et al., 2002; Sauer et al., 2012; Timmers et al., 2015). Nevertheless, from our experience in operating high pressure bioreactors, the enrichment of ANME under standard atmospheric pressure conditions is more cost effective, providing ease of reactor operation and avoiding explosion risks. Therefore, the enrichment of these methanotrophic *Archaea* can be done at atmospheric pressure conditions in order to investigate the ANME-3 metabolism and its potential cooperation (syntrophy) with a bacterial partner. After enriching ANME-3, further tests for the application of AOM-SR in treating sulfate containing industrial wastewater should be explored.

8.3   Anaerobic oxidation of methane in bioreactors

ANME enrichment has been a constant challenge for researchers performing AOM studies. Due to lack of a defined enrichment procedure and/or protocol for ANME, the explorations of ANME mechanisms take longer, sometimes up to 8-10 years. One typical example is mentioned here: almost 95% of the ANME-2 and *Desulfosarcinales* enriched consortia was achieved after cultivating the sediment from Isis mud volcano for more than 8 years (Milucka et al., 2012). Such long-term enrichments allow investigations on the ANME physiology and mechanisms and provide an insight on the novel AOM mechanisms as well as the role of alternate substrates for AOM. Therefore, testing of different bioreactor configurations could be potentially advantageous for ANME enrichment. Several researchers have tested batch, fed-batch, high-pressure and continuous-flow bioreactor systems for the enrichment of AOM microbial communities and ascertained the AOM rates and the role of different electron donors and electron acceptors. In Chapters 5, 6 and 7 of this thesis, we showed the ANME enrichment and its performance in three different bioreactor configurations, i.e. a biotrickling filter (BTF, Chapter 5), a membrane bioreactor (MBR, Chapter 6) and a high pressure bioreactor (HPB, Chapter 7).

During reactor startup, both the BTF and MBR were inoculated with the sediment slurry obtained from same location, i.e. Ginsburg mud volcano (Gulf of Cadiz, Spain) and operated under the same process conditions at ambient temperature and pressure with a continuous supply of methane (with 99.9% purity). Sulfate reduction with the simultaneous production of sulfide and consumption of methane was observed only after 150 days of bioreactor operation in both the BTF and MBR (Chapters 5 and 6). The key design considerations for both bioreactors were the biomass retention provided by the packing material (BTF) or the membrane (MBR) as well as the constant supply of methane aided in the proliferation of ANME. After the start up period of 150 days, both bioreactors showed a good performance and resilience capacity for AOM enrichment. Nevertheless, the sulfate reduction was slightly higher and faster in the BTF with a maximum volumetric sulfate reduction rate of 1.3 mM d^{-1} on day 280, whereas the maximum volumetric rate of sulfate reduction (0.5 mM d^{-1}) was obtained in the MBR after 380 days of bioreactor operation. One of the reasons for the faster AOM performance in the BTF compared to the MBR was the attachment of biomass in the porous polyurethane foam used as the packing material. In contrast to the packing material in the BTF, the biomass in the MBR experienced more stress due to continuous bubbling of methane and recirculation of artificial seawater medium, which might have possibly disturbed the biomass that was present at the bottom of the reactor (Meulepas et al., 2009).

HPB bioreactors with elevated methane partial pressure (up to 10 MPa) have been used previously by other researchers in order to enrich ANME and to achieve higher AOM coupled to sulfate reduction rates (Deusner et al., 2009; Zhang et al., 2010; Zhang et al., 2011b). In our study, during high pressure incubations of a highly enriched ANME-2 community, the incubation at *in situ* temperature (average temperature ~ 15°C) and 10 MPa pressure were found to be the optimum conditions for the active AOM-SR. A high population growth rate of ANME-2 was observed when Eckernförde Bay sediment was incubated at high pressure, i.e. 10 MPa pressure (Timmers et al., 2015). The ANME growth rate might be faster due to the higher negative Gibbs free energy at higher methane partial pressure. The affinity constant for methane of anaerobic methanotrophs from the Gulf of Cadiz sediment (i.e. the same sediment used in this study for the high pressure incubation) is around 37 mM, which is almost the total amount of dissolved methane at 30 MPa methane partial pressure (Zhang et al., 2010). Nevertheless, as discussed in Chapter 7 of this thesis, we showed the sulfate reducing AOM activity at 10 MPa incubation was comparatively higher than the incubations with relatively higher pressures, i.e. 20 MPa and 30 MPa. The highest AOM-SR rate (sulfide production rate of 0.18 mmol d^{-1} and total inorganic carbon production rate of 0.7 mmol d^{-1}) was achieved in HPB reactor incubated at 10 MPa. The results suggested that highest AOM-SR activity might be attributed to the origin of biomass, where the *in situ* pressure is also almost 10 MPa. Furthermore, we demonstrated that the activity of ANME-2 enrichment used in this study might be influenced by the availability of methane, albeit the pressures exceeding 20 MPa can also influence the cells of the ANME and sulfate reducing communities by disrupting the structure of several cellular components and functions, such as membrane fluidity, protein activity and structure (Jebbar et al., 2015), resulting in a retardation or decline of the activity.

8.4   Enrichment of ANME in different bioreactor configurations

In Chapters 5 and 6, the two distinct ANME phylotypes were cultivated majorly in the two bioreactor configurations (Figure 8.2)., i.e. in the BTF and the MBR. In the BTF, the relative abundance of ANME-1 was almost 4 times higher than ANME-2 as detected by 16S rRNA sequencing by Illumina Miseq method, whereas in the MBR, a dominancy of the ANME-2/ *Desulfosarcina* consortium was shown by catalyzed reporter deposition-fluorescence *in situ* hybridization (CARD-FISH) method.

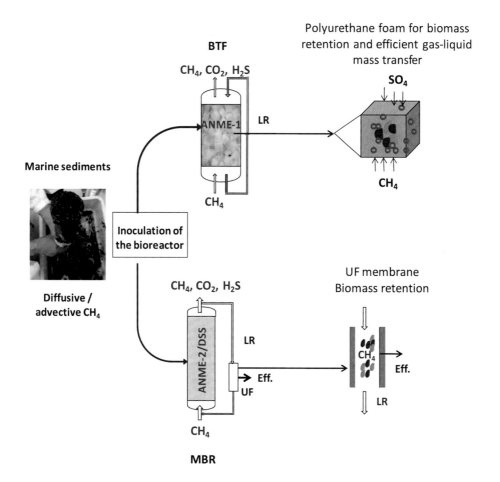

Figure 8.2 *Schematic representing the membrane bioreactor (MBR) and biotrickling filter type bioreactor (BTF) for the ANME enrichment and their characteristic feature to promote ANME growth. LR - liquid recirculation, UF - ultrafiltration, Eff - effluent.*

Both methodologies used have their own limitation, especially DNA based sequencing is biased by the PCR method and might be only an indirect indicator of the dominant functional community (Schirmer et al., 2015). The CARD-FISH analysis overcomes such PCR biases (Lloyd et al., 2013) and indicates the active microbial clades.

Despite these methodological biases that could not be avoided, our results indicated that one phylotype of ANME can be preferred over another depending on the bioreactor configuration. The shear force created in the suspended biomass might be one of the natural selection factors for ANME-2 dominancy in the MBR. From a marine sediment perspective, the majority of ANME-1 dominant habitats are the microbial mats and carbonate chimneys (Blumenberg et al., 2004; Michaelis et al., 2002), which provides a porous matrix for ANME proliferation. This porous natural matrix can harbor aggregates of AOM performing consortia (Marlow et al., 2014). ANME-1 and its growth rate appeared to be favored by the porous support material present in the BTF. The filter bed facilitated good gaseous mass transfer towing to which methane, having low solubility in water, can be captured and retained for longer period within the pores of the sponges.

Similar to the results obtained from enriched biomass of the MBR, enrichment of ANME-2 and *Desulfosarcinales* is reported in the majority of the long term incubations and continuous bioreactors performing AOM (Aoki et al., 2014; Meulepas et al., 2009; Timmers et al., 2015; Zhang et al., 2011b). In this study, ANME-2 and *Desulfosarcinales* were often closely associated to each other during CARD-FISH analyses which support the theory of potential syntrophy between ANME-2 and *Desulfosarcinales*. Co-occurrence of ANME-2 and *Desulfosarcinales* clades indicates their symbiotic relationship for performing AOM-SR and this phenomenon has been previously reported from several natural habitats (discussed in Chapter 6). However, the intermediates were not detected yet in previous *ex situ* studies on AOM. In this work, acetate formation was detected in the MBR, which could be possibly an intermediate from methane conversion. If the acetate formation was detected in the MBR, the question still remains whether the ANME, the SRB or any other clade were responsible for the pathway as well as the acetate formation. The acetate formation pattern in the MBR was also intermittent, which clearly suggests the formation and the consumption of acetate as a result of microbial metabolism. Recently, some of the *Desulfosarcinales* related subgroups obtained from marine hydrocarbon seep were found to be involved in the metabolism of other short chain hydrocarbons, such as butane, rather than methane (Kleindienst et al., 2014). In any bioreactor enrichment, diverse microbial metabolism and pathways can interact together leading to the formation of several intermediates. Further studies with ^{13}C labeled acetate in the enriched biomass and tracking of the substrate consumption pattern of ANME and associated SRB can provide a much clearer understanding on the mechanisms and pathways involved.

8.5   Future perspectives in ANME enrichments

An efficiently designed bioreactor for the enrichment of the microbial community responsible for AOM can be a scientific breakthrough for the further exploration of the ecophysiology of ANME and its potential biotechnological application. Mostly ANME enrichment studies have taken several years of bioreactor operation prior to achieve any active enrichments, i.e. the cultivation of ANME in a down-flow hanging sponge bioreactor that was operated for 5.5 years, in a continuous enrichment of ANME in a membrane bioreactor for 2.5 years (Meulepas et al., 2009) and a continuous cultivation of ANME more than 8 years (Milucka et al., 2012).

As the majority of ANME habitats are located in deep-sea environments with high methane partial pressures, simulation of the natural environment in the laboratory has shown advantageous for faster ANME growth and a relatively rapid bioreactor start-up period (Deusner et al., 2009; Timmers et al., 2015; Zhang et al., 2010; Zhang et al., 2011b). Nevertheless, reproducing *in situ* conditions in the laboratory and handling a high pressure system is quite challenging, especially for biotechnological and environmental engineering applications. In this thesis, we achieved the enrichment of both ANME-1 and ANME-2 in two different bioreactor designs operated under ambient conditions. The bioreactors start-up phase was comparatively shorter (150 days) than that of previous bioreactors that were tested under ambient conditions (~ 400 days) (Aoki et al., 2014; Meulepas et al., 2009). Based on our results, the BTF showed a higher rate of AOM dependent sulfate reduction and almost 50% ANME enrichment was achieved in ~ 250 days of bioreactor operation. Further optimization of the BTF configuration, testing different methane flow patterns, as well as enriching other sediments from diverse ANME habitats are recommended for future studies.

We have located the AOM site in coastal sediment from Marine Lake Grevelingen (the Netherlands) with the dominancy of ANME-3 (Bhattarai et al., 2017). Due to its similarity with *in situ* condition, i.e. ambient pressure conditions, enrichment of this anaerobic AOM community from the Marine Lake Grevelingen sediments in bioreactors will be easier and much efficient compared to enrichment of deep-sea sediments in laboratory scale bioreactors. In the current scenario of very limited knowledge on the ANME-3 ecophysiology, the enrichment of ANME-3 in bioreactors would be beneficial to understand the mechanism and physiology of ANME-3. Moreover, we have achieved enrichment of ANME-1 in the BTF and ANME-2 in the MBR. Based on the knowledge gained from this work, ANME could be enriched in different bioreactor configurations and thus enriched community can be further tested to understand its mechanisms, which seems promising advancement in the technical approach of AOM studies and way forward for its potential applications such as the removal of sulfate from wastewater using a cheap electron donor as methane. It is appealing to locate the ANME habitat in shallow marine environments, understand their key microbial processes and turnover and establish the links and roles played between different microbial communities when these sediments are transferred and cultivated in continuously operated bioreactors.

8.6 References

Aoki, M., Ehara, M., Saito, Y., Yoshioka, H., Miyazaki, M., Saito, Y., Miyashita, A., Kawakami, S., Yamaguchi, T., Ohashi, A., Nunoura, T., Takai, K., Imachi, H. 2014. A long-term cultivation of an anaerobic methane-oxidizing microbial community from deep-sea methane-seep sediment using a continuous-flow bioreactor. *PLoS ONE*, **9**(8), e105356.

Aquilina, A., Knab, N.J., Knittel, K., Kaur, G., Geissler, A., Kelly, S.P., Fossing, H., Boot, C.S., Parkes, R.J., Mills, R.A., Boetius, A., Lloyd, J.R., Pancost, R.D. 2010. Biomarker indicators for anaerobic oxidizers of methane in brackish-marine sediments with diffusive methane fluxes. *Org. Geochem.*, **41**(4), 414-426.

Bhattarai, S., Cassarini, C., Gonzalez-Gil, G., Egger, M., Slomp, C.P., Zhang, Y., Esposito, G., Lens, P.N.L. 2017. Anaerobic methane-oxidizing microbial community in a coastal marine sediment: anaerobic methanotrophy dominated by ANME-3. *Microb. Ecol.*, **74**(3), 608-622.

Biddle, J.F., Cardman, Z., Mendlovitz, H., Albert, D.B., Lloyd, K.G., Boetius, A., Teske, A. 2011. Anaerobic oxidation of methane at different temperature regimes in Guaymas Basin hydrothermal sediments. *ISME J.*, **6**(5), 1018-1031.

Blumenberg, M., Seifert, R., Reitner, J., Pape, T., Michaelis, W. 2004. Membrane lipid patterns typify distinct anaerobic methanotrophic consortia. *Proc. Natl. Acad. Sci. USA*, **101**(30), 11111-11116.

Boetius, A., Holler, T., Knittel, K., Felden, J., Wenzhöfer, F. 2009. The seabed as natural laboratory: lessons from uncultivated methanotrophs. in: Epstein, S.S. (Ed.), *Uncultivated Microorganisms*. Vol 10 of series Microbiology monographs, Springer Berlin Heidelberg, Germany, pp. 293-316.

Boetius, A., Ravenschlag, K., Schubert, C.J., Rickert, D., Widdel, F., Gieseke, A., Amann, R., Jørgensen, B.B., Witte, U., Pfannkuche, O. 2000. A marine microbial consortium apparently mediating anaerobic oxidation of methane. *Nature*, **407**, 623-626.

Boetius, A., Wenzhofer, F. 2013. Seafloor oxygen consumption fuelled by methane from cold seeps. *Nat. Geosci.*, **6**(9), 725-734.

Brazelton, W.J., Schrenk, M.O., Kelley, D.S., Baross, J.A. 2006. Methane-and sulfur-metabolizing microbial communities dominate the Lost City hydrothermal field ecosystem. *Appl. Environ. Microbiol.*, **72**(9), 6257-6270.

Dale, A.W., Aguilera, D., Regnier, P., Fossing, H., Knab, N., Jørgensen, B.B. 2008. Seasonal dynamics of the depth and rate of anaerobic oxidation of methane in Aarhus Bay (Denmark) sediments. *J. Mar. Res.*, **66**(1), 127-155.

Deusner, C., Meyer, V., Ferdelman, T. 2009. High-pressure systems for gas-phase free continuous incubation of enriched marine microbial communities performing anaerobic oxidation of methane. *Biotechnol. Bioeng.*, **105**(3), 524-533.

Heijs, S.K., Haese, R.R., Van der Wielen, P.W., Forney, L.J., Van Elsas, J.D. 2007. Use of 16S rRNA gene based clone libraries to assess microbial communities potentially involved in anaerobic methane oxidation in a Mediterranean cold seep. *Microb. Ecol.*, **53**(3), 384-398.

Jebbar, M., Franzetti, B., Girard, E., Oger, P. 2015. Microbial diversity and adaptation to high hydrostatic pressure in deep-sea hydrothermal vents prokaryotes. *Extremophiles*, **19**(4), 721-740.

Kleindienst, S., Herbst, F.-A., Stagars, M., von Netzer, F., von Bergen, M., Seifert, J., Peplies, J., Amann, R., Musat, F., Lueders, T. 2014. Diverse sulfate-reducing bacteria of the *Desulfosarcina/Desulfococcus* clade are the key alkane degraders at marine seeps. *ISME J.*, **8**(10), 2029-2044.

Lloyd, K.G., May, M.K., Kevorkian, R.T., Steen, A.D. 2013. Meta-analysis of quantification methods shows that archaea and bacteria have similar abundances in the subseafloor. *Appl. Environ. Microbiol.*, **79**(24), 7790-7799.

Losekann, T., Knittel, K., Nadalig, T., Fuchs, B., Niemann, H., Boetius, A., Amann, R. 2007. Diversity and abundance of aerobic and anaerobic methane oxidizers at the Haakon Mosby mud volcano, Barents Sea. *Appl. Environ. Microbiol.*, **73**(10), 3348-3362.

Marlow, J.J., Steele, J.A., Ziebis, W., Thurber, A.R., Levin, L.A., Orphan, V.J. 2014. Carbonate-hosted methanotrophy represents an unrecognized methane sink in the deep sea. *Nat. Commun.*, **5**(5094), 1-12.

Meulepas, R.J., Stams, A.J., Lens, P.N. 2010. Biotechnological aspects of sulfate reduction with methane as electron donor. *Rev. Environ. Sci. Biotechnol.*, **9**(1), 59-78.

Meulepas, R.J.W., Jagersma, C.G., Gieteling, J., Buisman, C.J.N., Stams, A.J.M., Lens, P.N.L. 2009. Enrichment of anaerobic methanotrophs in sulfate-reducing membrane bioreactors. *Biotechnol. Bioeng.*, **104**(3), 458-470.

Michaelis, W., Seifert, R., Nauhaus, K., Treude, T., Thiel, V., Blumenberg, M., Knittel, K., Gieseke, A., Peterknecht, K., Pape, T., Boetius, A., Amann, R., Jørgensen, B.B., Widdel, F., Peckmann, J., Pimenov, N.V., Gulin, M.B. 2002. Microbial reefs in the Black Sea fueled by anaerobic oxidation of methane. *Science*, **297**, 1013-1015.

Milucka, J., Ferdelman, T.G., Polerecky, L., Franzke, D., Wegener, G., Schmid, M., Lieberwirth, I., Wagner, M., Widdel, F., Kuypers, M.M.M. 2012. Zero-valent sulphur is a key intermediate in marine methane oxidation. *Nature*, **491**(7425), 541-546.

Nauhaus, K., Boetius, A., Kruger, M., Widdel, F. 2002. *In vitro* demonstration of anaerobic oxidation of methane coupled to sulphate reduction in sediment from a marine gas hydrate area. *Environ. Microbiol.*, **4**(5), 296-305.

Niemann, H., Losekann, T., de Beer, D., Elvert, M., Nadalig, T., Knittel, K., Amann, R., Sauter, E.J., Schluter, M., Klages, M., Foucher, J.P., Boetius, A. 2006. Novel microbial communities of the Haakon Mosby mud volcano and their role as a methane sink. *Nature*, **443**(7113), 854-858.

Oni, O.E., Miyatake, T., Kasten, S., Richter-Heitmann, T., Fischer, D., Wagenknecht, L., Ksenofontov, V., Kulkarni, A., Blumers, M., Shylin, S. 2015. Distinct microbial populations are tightly linked to the profile of dissolved iron in the methanic sediments of the Helgoland mud area, North Sea. *Front. Microbiol.*, **6**(365), 1-15.

Orphan, V.J., Ussler, W., Naehr, T.H., House, C.H., Hinrichs, K.U., Paull, C.K. 2004. Geological, geochemical, and microbiological heterogeneity of the seafloor around methane vents in the Eel River Basin, offshore California. *Chem. Geol.*, **205**(3), 265-289.

Pachiadaki, M., G. , Lykousis, V., Stefanou, E., G., Kormas, K., A. 2010. Prokaryotic community structure and diversity in the sediments of an active submarine mud volcano (Kazan mud volcano, East Mediterranean Sea). *FEMS. Microbiol. Ecol.*, **72**(3), 429-444.

Ruff, S.E., Kuhfuss, H., Wegener, G., Lott, C., Ramette, A., Wiedling, J., Knittel, K., Weber, M. 2016. Methane seep in shallow-water permeable sediment harbors high diversity of anaerobic methanotrophic communities, Elba, Italy. *Front. Microbiol.*, **7**(374), 1-20.

Sauer, P., Glombitza, C., Kallmeyer, J. 2012. A system for incubations at high gas partial pressure. *Front. Microbiol.*, **3**(25), 225-233.

Schirmer, M., Ijaz, U.Z., D'Amore, R., Hall, N., Sloan, W.T., Quince, C. 2015. Insight into biases and sequencing errors for amplicon sequencing with the Illumina MiSeq platform. *Nucl. Acids Res.*, **(43)**6, e37, doi: 10.1093/nar/gku1341.

Sivan, O., Schrag, D.P., Murray, R.W. 2007. Rates of methanogenesis and methanotrophy in deep-sea sediments. *Geobiology*, **5**(2), 141-151.

Thornburg, C.C., Zabriskie, T.M., McPhail, K.L. 2010. Deep-Sea hydrothermal vents: potential hot spots for natural products discovery? *J. Nat. Prod.*, **73**(3), 489-499.

Thurber, A., Sweetman, A., Narayanaswamy, B., Jones, D., Ingels, J., Hansman, R. 2014. Ecosystem function and services provided by the deep sea. *Biogeosciences*, **11**(14), 3941-3963.

Timmers, P.H., Gieteling, J., Widjaja-Greefkes, H.A., Plugge, C.M., Stams, A.J., Lens, P.N., Meulepas, R.J. 2015. Growth of anaerobic methane-oxidizing archaea and sulfate-reducing bacteria in a high-pressure membrane capsule bioreactor. *Appl. Environ. Microbiol.*, **81**(4), 1286-1296.

Treude, T., Krüger, M., Boetius, A., Jørgensen, B.B. 2005. Environmental control on anaerobic oxidation of methane in the gassy sediments of Eckernförde Bay (German Baltic). *Limnol. Oceanogr.*, **50**(6), 1771-1786.

Ussler III, W., Preston, C., Tavormina, P., Pargett, D., Jensen, S., Roman, B., Marin III, R., Shah, S.R., Girguis, P.R., Birch, J.M. and Orphan, V., 2013. Autonomous application of quantitative PCR in the deep sea: *in situ* surveys of aerobic methanotrophs using the deep-sea environmental sample processor. *Environ. Sci. Technol.*, **47**(16), 9339-9346.

Zhang, Y., Arends, J.B., Van de Wiele, T., Boon, N. 2011a. Bioreactor technology in marine microbiology: from design to future application. *Biotechnol. Adv.*, **29**(3), 312-321.

Zhang, Y., Maignien, L., Zhao, X., Wang, F., Boon, N. 2011b. Enrichment of a microbial community performing anaerobic oxidation of methane in a continuous high-pressure bioreactor. *BMC Microbiol.*, **11**(137), 1-8.

Zhang, Y., Henriet, J.-P., Bursens, J., Boon, N. 2010. Stimulation of *in vitro* anaerobic oxidation of methane rate in a continuous high-pressure bioreactor. *Bioresour. Technol.*, **101**(9), 3132-3138.

Biography

Susma Bhattarai Gautam was born in Nepal and did her masters (MSc) in Environmental Science from UNESCO-IHE, Institute of Water Education, Delft, the Netherlands with the specialization in Environmental Science and Technology. During her MSc, She worked on a research project from EAWAG, Switzerland which aimed to characterize microbial community of Lake Kivu sediment, East Africa. Prior to that, she did her bachelor (BSc) and MSc in Environmental Science with the specialization in water resource management from Tribuvan University, Nepal. She had also worked as an environmentalist in hydropower projects of Nepal and her major responsibility was to carry out environmental impact assessment of potential hydropower projects. Susma got admitted in Erasmus Mundus Joint Doctorate Programme Environmental Technologies for Contaminated Solids, Soils and Sediments (ETeCoS^3) and started as PhD fellow at UNESCO-IHE from March, 2013. As a part of the programme, she also carried her research work at Shanghai Jiao Tong University, China and University of Cassino and Southern Lazio, Italy. Her research was mainly focused on the study of anaerobic oxidation of methane, enrichment of anaerobic methanotrophic communities which perform anaerobic oxidation of methane in bioreactors and understand their ecophysiology.

Publications

Bhattarai, S., Cassarini, C., Rene, E.R., Y. Zhang, Esposito, G., Lens, P.N.L. 2018. Enrichment of sulfate reducing anaerobic methane oxidizing community dominated by ANME-1 from Ginsburg Mud Volcano (Gulf of Cadiz) sediment in a biotrickling filter. *Bioresource Technology, 259* (2018), 433-441

Bhattarai, S., Cassarini, C., Rene, E.R., Esposito, G., Lens, P.N.L. 2018. Enrichment of ANME-2 dominated anaerobic methanotrophy from cold seep sediment in an external ultrafiltration membrane bioreactor. *Life Science Engineering,0,* 1-11.

Bhattarai S, Cassarini C, Gonzalez-Gil G., Egger M., Slomp C.P., Zhang Y, Esposito G. and Lens P N. L. 2017. Anaerobic methane-oxidizing microbial community in a coastal marine sediment: anaerobic methanotrophy dominated by ANME-3, *Microbial Ecology Journal*, 74:608-622

Bhattarai S, Cassarini C., Naangmenyele Z., Rene E. R., Gonzalez-Gil G., Esposito G. and Lens P N. L. 2017. Microbial sulfate reducing activities in anoxic sediment from Marine Lake Grevelingen: screening of electron donors and acceptors, *Limnology*, 1-11

Bhattarai S, Ross K-A, Schmid M, Anselmetti F, and Bürgmann H. 2012. Local conditions structure unique archaeal communities in the anoxic sediments of meromictic Lake Kivu, *Microbial Ecology Journal*, 64:291-310

Cassarini C., Rene E. R., **Bhattarai S.**, Esposito G., and Lens P. N. L. 2017. Anaerobic oxidation of methane coupled to thiosulfate reduction in a biotrickling filter, *Bioresource Technology*, 240:214-222

Conferences

Sulfate reduction by marine sediments hosting anaerobic oxidation of methane. **Bhattarai S**, Cassarini C, Gonzalez-Gil G., Egger M., Slomp C.P., Zhang Y, P. N. L. Lens: The 4th international conference on Research Frontiers in Chancogen Science and Technology, G16, Delft , the Netherlands, May 28-29 (oral presentation and conference proceeding)

Anaerobic oxidation of methane in surface sediments of Marine Lake Grevelingen, North Sea. **Bhattarai S**, Cassarini C, Gonzalez-Gil G., Egger M., Slomp C.P., Zhang Y, P.N.L. Lens: The 6th Congress of European Microbiologists, FEMS 2011, Maastricht, the Netherlands, June 7-11, 2015 (Poster)

Methane Cycling and Microbial Population in Lake Kivu. H. Bürgmann, N. Pasche, **S. Bhattarai**, K.A. Ross, F. Vazquez, C. J. Schubert, A. Wüest, M. Schmid. Poster: The 4th Congress of European Microbiologists, FEMS 2011, Geneva, Switzerland, June 26-30, 2011 (Poster)

Netherlands Research School for the
Socio-Economic and Natural Sciences of the Environment

D I P L O M A

For specialised PhD training

The Netherlands Research School for the
Socio-Economic and Natural Sciences of the Environment
(SENSE) declares that

Susma Bhattarai Gautam

born on 13 December 1980 in Jhapa, Nepal

has successfully fulfilled all requirements of the
Educational Programme of SENSE.

Cassino, 16 December 2016

the Chairman of the SENSE board

Prof. dr. Huub Rijnaarts

the SENSE Director of Education

Dr. Ad van Dommelen

The SENSE Research School has been accredited by the Royal Netherlands Academy of Arts and Sciences (KNAW)

K O N I N K L I J K E   N E D E R L A N D S E
A K A D E M I E   V A N   W E T E N S C H A P P E N

The SENSE Research School declares that **Ms Susma Bhattarai Gautam** has successfully
fulfilled all requirements of the Educational PhD Programme of SENSE with a
work load of 38.1 EC, including the following activities:

SENSE PhD Courses

o Environmental research in context (2013)
o Research in context activity: 'Co-organizing the programme and abstract book of the 4th
   International Conference on Research Frontiers in Chalcogen Cycle Science & Technology,
   Delft' (2015)

Other PhD and Advanced MSc Courses

o Contaminated sediments, characterization and remediation, UNESCO-IHE, The Netherlands
   (2013)
o Anearobic waste treatment, University of Cassino and Southern Lazio, Italy (2014)
o Biological treatment of solid waste, University of Cassino and Southern Lazio, Italy (2014)
o Anaerobic waste water treatment, UNESCO-IHE, The Netherlands (2014)
o Contaminated soils, Université Paris-Est Marne-la-Vallée, France (2015)

External training at a foreign research institute

o Research collaboration, bioinformatics software training, and training for the use of
   Polymerase Chain Reaction (PCR) and Q-PCR equipment, Shanghai Jiao Tong University,
   China (2014-2015)

Management and Didactic Skills Training

o Organisational support for the ETeCoS3 summer school on 'Contaminated sediments,
   characterization and remediation' given at UNESCO-IHE (2013)
o Supervising MSc student with thesis entitled 'Anaerobic oxidation of methane and sulphate
   reduction by marine sediments in the presence of different electron donors and acceptors'
   (2013-2014)
o Assisting lab-practicals of the MSc course 'Microbiology' (2013-2014)
o Assisting lectures and lab-practicals of the MSc course 'Aquatic ecosystems' (2014)

Selection of Oral and Poster presentations

o *Sulphate reduction by marine sediments hosting anaerobic oxidation of methane.*
   4th International Conference on Research Frontiers in Chalcogen Cycle Science & Technology,
   28-29 May 2015, Delft, The Netherlands
o *Anaerobic oxidation of methane in surface sediments of marine lake Grevelingen* [poster].
   6th Congress of European Microbiologists (FEMS), 7-11 June 2015, Maastricht, The
   Netherlands

SENSE Coordinator PhD Education

Dr. ing. Monique Gulickx

Printed and bound by CPI Group (UK) Ltd, Croydon, CR0 4YY

21/10/2024

01777112-0015